INTERNATIONAL SERIES OF MONOGRAPHS ON
PURE AND APPLIED BIOLOGY

Division: **ZOOLOGY**

GENERAL EDITOR: G. A. KERKUT

VOLUME 4

IMPLICATIONS OF EVOLUTION

IMPLICATIONS OF EVOLUTION

By

G. A. KERKUT

M.A., PH.D.

Department of Physiology and Biochemistry
The University of Southampton

PERGAMON PRESS

NEW YORK · OXFORD · LONDON · PARIS

1960

PERGAMON PRESS INC.
122 East 55th Street, New York 22, N.Y.
P.O. Box 47715, Los Angeles, California

PERGAMON PRESS LTD.
Headington Hill Hall, Oxford
4 & 5 Fitzroy Square, London W.1

PERGAMON PRESS S.A.R.L.
24 Rue des Écoles, Paris Ve

PERGAMON PRESS G.m.b.H.
Kaiserstrasse 75, Frankfurt am Main

Library of Congress Card No. 60–9644

MADE AND PRINTED IN GREAT BRITAIN BY
THE GARDEN CITY PRESS LIMITED
LETCHWORTH, HERTFORDSHIRE

CONTENTS

CONTENTS

PREFACE

THERE are many books about Evolution so perhaps in this preface I should state what this book is *not* about. It is not concerned with the mechanism of speciation, the evolution of dominance, the relationship of enzymatic adaptation to the inheritance of acquired characteristics or the probability that Natural Selection can bring about a pandemic of rodents in $n + 1$ years. Instead the present book is concerned with an examination of certain basic assumptions and implications that have become involved in the present-day concept of the evolutionary relationships within the animal kingdom. The majority of books on Evolution either blatantly treat these assumptions as part of an old (and concluded) historic argument or else they avoid discussing the assumptions and instead deal with the more scientific and mathematical parts of Evolution.

If one tries to question this avoiding reaction, the protagonists round on one and say in an accusing tone of voice, " Don't you believe in the Theory of Organic Evolution? What better theory have you got to offer? "

May I here humbly state as part of my biological *credo* that I believe that the theory of Evolution as presented by orthodox evolutionists is in many ways a satisfying explanation of some of the evidence. At the same time I think that the attempt to explain all living forms in terms of an evolution *from a unique source*, though a brave and valid attempt, is one that is premature and not satisfactorily supported by present-day evidence. It may in fact be shown ultimately to be the correct explanation, but the supporting evidence remains to be discovered. We can, if we like, believe that such an evolutionary system has taken place, but I for one do not think that " it has been proven beyond all reasonable doubt." In the pages of the book that follow I shall present evidence for the point of view that there are many discrete groups of animals and that we do not know how they have evolved nor how they are interrelated. It is possible that they might have evolved quite independently from discrete and separate sources. There are only a limited number of chemical elements that are capable of forming stable polymerisation compounds and it is not at all surprising that the same compounds have been formed on several occasions. Quite complex materials such as carbohydrates, peptides and even nucleic acids can be formed by irradiating water containing simple salts and gases.

It may be suggested that the problem we are examining here, namely that of the evolution and interrelationship of the basic living stocks is a major problem and one that will test the strength and ability of many hundreds of research workers. If this book merely indicates to some of the readers that certain lines of thought are still open to examination, then I shall consider that it has done its allotted task.

There is, however, a second point that I should like to make, and this concerns not factual material but an attitude of mind. It is very depressing to find that many subjects are becoming encased in scientific dogmatism. The basic information is frequently overlooked or ignored and opinions become repeated so often and so loudly that they take on the tone of Laws. Although it does take a considerable amount of time, it is essential that the basic information is frequently re-examined and the conclusions analysed. From time to time one must stop and attempt to think things out for oneself instead of just accepting the most widely quoted viewpoint. I have dealt with this attitude in the introductory chapter of this book, though I hope that the moral does not end there but instead runs through the rest of the book as well.

It is a pleasure to acknowledge the kind help and assistance that various colleagues have given me during the writing of this work. Many of them have read through parts of the book or offered advice on various points. I have profited greatly from their counsel, though of course I bear full responsibility for all the statements and errors. In particular I should like to thank Professor E. Baldwin, Drs. M. S. Laverack, K. A. Munday, S. Smith, Miss D. Wisden, Messrs. Robert Walker, Edward Munn, and Richard Solly for their help and forbearance.

ACKNOWLEDGEMENTS

I am grateful to the following authors and publishers for permission to use figures and to quote from their publications:

Professor G. Schramm for kindly providing the photograph from which Fig. 2 was taken.

Dr. S. Brenner and Dr. R. W. Horne for kindly providing the photograph from which Fig. 3 was taken.

Academic Press for permission to quote from the article by H. A. Krebs in *Chemical Pathways in Metabolism*. Vol. 1, edited by D. M. Greenberg.

Allen and Unwin for permission to quote from the article by G. R. de Beer in " The evolution of the Metazoa " from the book *Evolution as a Process* edited by J. Huxley, A. C. Hardy and E. B. Ford.

Cambridge University Press for permission to take the table printed on page 112 from *Comparative Biochemistry* by E. Baldwin; the table on page 2 from *The Cambridge University Handbook*; and the quotations from *Biochemistry and Morphogenesis* by J. Needham, and *Growth and Form* by W. D'Arcy Thompson.

Masson et Cie for permission to reproduce Figs. 7, 8, 9, 11, 12, 15, and 16, from *Traité de Zoologie* edited by P. P. Grassé.

McGraw Hill for permission to reproduce Figs. 25 and 27, taken from *The Invertebrates* by L. H. Hyman, and also for permission to quote from this work.

Oxford University Press for permission to quote from E. Radl's *History of Biological Theories* and to adapt Fig. 43 from *The Horses* by G. G. Simpson.

Charles C. Thomas for permission to quote from a paper by H. A. Krebs published in the *Harvey Lectures*.

John Wiley for permission to take the table on page 20 from *General Biochemistry* by J. S. Fruton and S. Simmons.

x ACKNOWLEDGEMENTS

I should also like to thank the editors of the following journals for permission to copy and reproduce material. *Archive de Zoologie General et Experimentale; Biochemistry, Biophysics Acta; Biochemical Journal; Journal of Bacteriology; Journal of Experimental Biology; Proceedings of the Royal Society; Quarterly Review of Biology; Systematic Zoology* and *Zoologiska Bidrag.*

CHAPTER 1

INTRODUCTION

THROUGHOUT the Dark and Middle Ages, Learning was under the aegis of the Church. Except for useful subjects such as Medicine and perhaps Law, the university students were concerned with material that would either make the student a useful priest or else a person useful to priests.

The hold that the Church has had on the universities has been but slowly relinquished over the years. Until 1871 it was the custom for the majority of dons at Cambridge to be ordained before they could carry out any of the duties in college. This did not always mean that the prospective Fellow had to make a careful study of theology. Thus the Fellows of some colleges had the right of becoming ordained in their own chapel as soon as they were elected to a Fellowship without having to undergo any arduous extra study. This special sanction was taken away from them in 1852 and from then on they had to become ordained in the normal manner.

The Fellows besides being compulsorily ordained also had to live under an enforced celibacy. Should they wish to enjoy the varied pleasures of married life they had in turn to relinquish their college Fellowships. The married clergyman then left Cambridge and usually took up one of the livings that were in the gift of his college. This had its own compensations; those scholars who had swallowed their intellectual goat in their youth, instead of being forced to eke it out to various undergraduates for the rest of their lives, could leave Cambridge and take up a rich living in the outside world. This made more room available at the university for the younger man, who did not then merely have to wait for his older colleagues to die.

The hold of the Church on the university continued in many ways. The undergraduates coming up to Cambridge until 1852 had to be communicants of the Church of England, and the undergraduate coming up in, say, 1910 had to satisfy his examiners not only in his knowledge of classical languages but also had to show that he had some knowledge of Archdeacon William Paley's book on *Evidences of Christianity*. The latter examination was in force till 1927, when it was brought to the notice of the university authorities that many undergraduates did not in fact read Paley's *Evidences* but instead studied a little crib of them. Many of the more sceptical dons in the university were in favour of retaining the examination and ensuring that all undergraduates should be made to study Paley's *Evidences* most carefully, " For in this way," they said, " the student will be forced to realise just how weak the evidence in favour of Christianity really is." This argument was not upheld and in 1927 another piece of tradition was abandoned.

Many present-day undergraduates seem to imagine that the various subjects they study have existed as such, if not for eternity, then at least from time immemorial. They are surprised to learn that many of the chairs and examinations only came into existence over the last half-century. In the table below I have selected a few of the dates at which various chairs became established at Cambridge. It will be seen that the subjects of Theology and Medicine are very ancient whilst German, French and English are relatively modern.

Establishment of Chairs at Cambridge

1502	2 Chairs of Divinity
1540	Civil Law, Physic, Hebrew, Greek
1634	Arabic
1683	Moral Philosophy
1684	Philosophy
1702	Organic Chemistry
1704	Astronomy
1707	Anatomy
1724	History
1727	Botany
1866	Zoology
1869	Fine Art
1909	German

1911   English Literature
1919   French
1937   Geography
1938   Education

(This is only a selection from the complete list.)

You may ask, " What has all this got to do with evolution? "
It is my thesis that many of the Church's worst features are still
left embedded in present-day studies. Thus the serious under-
graduate of the previous centuries was brought up on a theological
diet from which he would learn to have faith and to quote authorities when he was in doubt. Intelligent understanding was the
last thing required. The undergraduate of today is just as bad;
he is still the same opinion-swallowing grub. He will gladly
devour opinions and views that he does not properly understand
in the hope that he may later regurgitate them during one of his
examinations. Regardless of his subject, be it Engineering,
Physics, English or Biology, he will have faith in theories that he
only dimly follows and will call upon various authorities to
support what he does not understand. In this he differs not one
bit from the irrational theology student of the bygone age who
would mumble his dogma and hurry through his studies in order
to reach the peace and plenty of the comfortable living in the world
outside. But what is worse, the present-day student *claims* to be
different from his predecessor in that he thinks scientifically and
despises dogma, and when challenged he says in defence, "After
all, one has to accept something, or else it takes a very long time to
get anywhere."

Well, let us see the present-day student " getting somewhere."
For some years now I have tutored undergraduates on various
aspects of Biology. It is quite common during the course of
conversation to ask the student if he knows the evidence for
Evolution. This usually evokes a faintly superior smile at the
simplicity of the question, since it is an old war-horse set in count-
less examinations. " Well, sir, there is the evidence from
palaeontology, comparative anatomy, embryology, systematics and
geographical distributions," the student will say in a nursery-
rhyme jargon, sometimes even ticking off the words on his fingers.
He would then sit and look fairly complacent and wait for a more

2—IOE

difficult question to follow, such as the nature of the evidence for
Natural Selection. Instead I would continue on with Evolution.

"Do you think that the Evolutionary Theory is the best
explanation yet advanced to explain animal interrelationships?"
I would ask.

"Why, of course, sir," would be the reply in some amazement
at my question. "There is nothing else, except for the religious
explanation held by some Fundamentalist Christians, and I
gather, sir, that these views are no longer held by the more
up-to-date Churchmen."

"So," I would continue, "you believe in Evolution because
there is no other theory?"

"Oh, no, sir," would be the reply, "I believe in it because of
the evidence I just mentioned."

"Have you read any book on the evidence for Evolution?"
I would ask.

"Yes, sir," and here he would mention the names of authors
of a popular school textbook, "and of course, sir, there is that
book by Darwin, *The Origin of Species*."

"Have you read this book?" I asked.

"Well, not all through, sir."

"About how much?"

"The first part, sir."

"The first fifty pages?"

"Yes, sir, about that much; maybe a bit less."

"I see, and that has given you your firm understanding of
Evolution?"

"Yes, sir."

"Well, now, if you really understand an argument you will
be able to indicate to me not only the points in favour of the
argument but also the most telling points against it."

"I suppose so, sir."

"Good. Please tell me, then, some of the evidence against the
theory of Evolution."

"Against what, sir?"

"The theory of Evolution."

"But there isn't any, sir."

Here the conversation would take on a more strained atmosphere.
The student would look at me as if I was playing a very unfair

game. It would be clearly quite against the rules to ask for evidence against a theory when he had learnt up everything in favour of the theory. He also would take it rather badly when I suggest that he is not being very scientific in his outlook if he swallows the latest scientific dogma and, when questioned, just repeats parrot fashion the views of the current Archbishop of Evolution. In fact he would be behaving like certain of those religious students he affects to despise. He would be taking on faith what he could not intellectually understand and when questioned would appeal to authority, the authority of a " good book " which in this case was *The Origin of Species*. (It is interesting to note that many of these widely quoted books are read by title only. Three of such that come to mind are the Bible, *The Origin of Species* and *Das Kapital*.)

I would then suggest that the student should go away and read the evidence for and against Evolution and present it as an essay. A week would pass and the same student would appear armed with an essay on the evidence for Evolution. The essay would usually be well done, since the student might have realised that I should be tough to convince. When the essay had been read and the question concerning the evidence against Evolution came up, the student would give a rather pained smile. " Well, sir, I looked up various books but could not find anything in the scientific books against Evolution. I did not think you would want a religious argument." " No, you were quite correct. I want a scientific argument against Evolution." " Well, sir, there does not seem to be one and that in itself is a piece of evidence in favour of the Evolutionary Theory."

At this piece of logic the student would sit back and feel that he had come out on top. After all, I had merely been questioning him whilst he had produced information.

I would then indicate to him that the theory of Evolution was of considerable antiquity and would mention that he might have looked at the book by Radl, *The History of Biological Theories*. Having made sure that the student had noted the book down for future reference I would proceed as follows.

CHAPTER 2

BASIC ASSUMPTIONS

BEFORE one can decide that the theory of Evolution is the best explanation of the present-day range of forms of living material one should examine all the implications that such a theory may hold. Too often the theory is applied to, say, the development of the horse and then because it is held to be applicable there it is extended to the rest of the animal kingdom with little or no further evidence.

There are, however, seven basic assumptions that are often not mentioned during discussions of Evolution. Many evolutionists ignore the first six assumptions and only consider the seventh. These are as follows.

(1) The first assumption is that non-living things gave rise to living material, i.e. spontaneous generation occurred.

(2) The second assumption is that spontaneous generation occurred only once.

The other assumptions all follow from the second one.

(3) The third assumption is that viruses, bacteria, plants and animals are all interrelated.

(4) The fourth assumption is that the Protozoa gave rise to the Metazoa.

(5) The fifth assumption is that the various invertebrate phyla are interrelated.

(6) The sixth assumption is that the invertebrates gave rise to the vertebrates.

(7) The seventh assumption is that within the vertebrates the fish gave rise to the amphibia, the amphibia to the reptiles, and the reptiles to the birds and mammals. Sometimes this is expressed in other words, i.e. that the modern amphibia and reptiles had a common ancestral stock, and so on.

For the initial purposes of this discussion on Evolution I shall consider that the supporters of the theory of Evolution hold that all these seven assumptions are valid, and that these assumptions form the " General Theory of Evolution."

The first point that I should like to make is that these seven assumptions by their nature are not capable of experimental verification. They assume that a certain series of events has occurred in the past. Thus though it may be possible to mimic some of these events under present-day conditions, this does not mean that these events must therefore have taken place in the past. All that it shows is that it is *possible* for such a change to take place. Thus to change a present-day reptile into a mammal, though of great interest, would not show the way in which the mammals did arise. Unfortunately we cannot bring about even this change; instead we have to depend upon limited circumstantial evidence for our assumptions, and it is now my intention to discuss the nature of this evidence.

Non-living into living (Biogenesis)

This is one of the oldest problems to puzzle man. Is it possible for non-living material simply to be turned into living material or is some extra " vital " force necessary? It is reasonably clear that living bodies in many ways use systems similar to those present in the non-living world. One of the first barricades appeared to fall to Wöhler, when he showed by his synthesis of urea that there was no very clear distinction between organic chemicals and non-organic chemicals. Within recent years we have been able to devise systems in which the irradiation of a mixture containing water, carbon dioxide and ammonia brings about the formation of amino-acids, simple peptides, and carbohydrates. However, proteins and nucleoproteins have not yet been synthesised under such conditions and these latter compounds appear to be of great importance in the development and maintenance of life. One imagines that the synthesis of these substances will merely be a matter of time and application, but it will be useful to distinguish the two different methods of achieving their synthesis. The first is to try to synthesise them under conditions in which we imagine that living things first occurred, i.e. to irradiate simple solutions and hope that proteins and nucleoproteins will form by random combination. This would mimic the conditions under which we believe life originated. The second method is to use specialised chemical and physical techniques to synthesise proteins and nucleoproteins, and having synthesised them, then to place them in their correct structural relationship. In this way, the combination of synthetic proteins, nucleic acids, lipids and carbohydrates might lead to the formation of a simple virus-like compound that could reproduce in living cells. The next stage would be the development of an artificial solution to maintain the artificial virus. With these steps accomplished we should have learnt a great deal about the processes taking place in the living body and no doubt we should have discovered new rules for physics and chemistry, but we could not say from our experiments that the living material in the universe arose in this way. The results would show that living matter can arise by synthetic methods devised in the laboratory, but it would still be possible that there were other methods by which life actually arose in the universe. For a full discussion of the origin of life one should consult the following articles: Oparin (1957); Bernal (1954); Pringle (1954); Pirie (1954); Haldane (1954).

Life arose only once

The assumption that life arose only once and that therefore all living things are interrelated is a useful assumption in that it provides a simple working basis for experimental procedure. But because a concept is useful it does not mean that it is necessarily correct. The experimental basis for this concept in particular is not as definite and as conclusive as many modern texts would have us believe.

Biochemical evidence.   Biochemists and comparative physiologists usually assume that all protoplasm, no matter where it is found, has the same fundamental biochemical and biophysical processes taking place in it. But even an elementary study of the situation shows that there are often many different ways of carrying out a simple process in the animal kingdom. One well-known example is that of carrying oxygen in solution; various substances such as haemoglobin, haemocyanin, haemerythrin and chlorocruorin are known to be capable of combining with oxygen. But the common possession of a specific blood pigment does not indicate any close phylogenetic relationship. Thus though many Crustacea have haemocyanin no biochemist or physiologist would suggest taking *Daphnia* out of the Crustacea because it possesses haemoglobin. The role of blood pigments has been much studied and in particular we accept the varied way in which they are distributed throughout the animal kingdom. Even here we do not always know their function; thus we find haemoglobin in the root nodules of leguminous plants (Keilin and Wang 1945), where its precise function is as yet not known. Plants, however, seem capable of synthesising many substances that are often regarded as " mammalian " compounds. Nettle stings contain acetyl choline, 5-hydroxytryptamine and histamine, and it is probable that these have been independently developed by the higher plants.

There are in the world but some ninety elements, and of these only a few such as carbon, nitrogen, oxygen, hydrogen, phosphorus and sulphur appear capable of forming natural monomers and polymers. It is therefore not surprising that these elements are united to form compounds such as citric acid or 5-hydroxy-tryptamine in widely separated plants and animals. Such a synthesis might have occurred independently on many occasions by trial and error. It should be remembered that there is no Patent Law in the natural world, and though one can simplify the situation by use of William of Occam's razor, the careless use of such a weapon can at times be suicidal.

Our ignorance is even greater in other biochemical fields, yet it is often stated that all protoplasm shows the same fundamental biochemical systems. The most quoted example is the way in which protoplasm oxidises carbohydrates to liberate energy. This release of energy is obtained through two biochemical cycles, the glycolysis cycle (Embden–Meyerhof) and the tricarboxylic acid cycle (Krebs). Many of the chemicals present in these two cycles have been found in bacteria, protozoa, plants, lower metazoa, birds and mammals, and because some of the ingredients are present it is assumed that the whole system is present. The argument then runs that because the system is very complex, it would be too much to expect that each group developed this complex system independently and so protoplasm everywhere must have had a common origin.

Krebs in 1948 discussed the universality of the tricarboxylic cycle in cells and tissues. He stated, " there is no doubt that yeast cells can synthesise succinate in the presence of glucose, and citrate in the presence of acetate, but none of the strains of baker's and brewer's yeast tested at the Sheffield laboratory was found capable of oxidising succinic or citric acids at a significant rate under whatever conditions these substances were tested." In 1954 he was of much the same opinion: " thus all the enzyme systems required for the tricarboxylic cycle are present in yeast cells and there can be no doubt that the cycle *can* take place. . . . However, these findings are not decisive evidence for the assumption that the cycle is the main terminal respiratory process in yeasts. . . . In many other organisms another terminal oxidation mechanism seems to play a major role. Its nature is unknown in the case of yeast. It may be a dicarboxylic acid cycle in certain bacteria." It now appears that the events that suggested the existence of a dicarboxylic acid cycle in bacteria may be better explained in terms of a divergence from the tricarboxylic acid cycle (Kornberg 1958). The system of terminal oxidation in yeasts is still obscure.

In effect, then, the situation in bacteria, yeasts, plants and the lower animals is not as simple or clear cut as might be imagined. There is more than one pathway for the breakdown of carbohydrates, and the glycolysis cycle and the citric acid cycle are but two of many that are in the process of elucidation. Thus recently a hexose monophosphate shunt has been described as an alternative method by which bacteria and many animal tissues break down glucose. This, combined with the possibility of an alternative terminal oxidation system, enables us to postulate two more systems that may be active in tissue metabolism. This view is supported by Cohen (1955a), who in his account of alternative pathways in carbohydrate metabolism states, " the time is past when we uncritically ascribe phenomena in carbohydrate metabolism to variations in the Embden–Meyerhof scheme." Cohen (1955b) also suggests that at least six major pathways for glucose metabolism are known and several may exist simultaneously in the same organism.

When one considers the various animals and bacteria that have been studied, it becomes quite clear that what we have so far

FIG. 1a. Photograph of material found in section of meteorite. Such sections may show shapes that resemble living forms. It is not clear if these in fact were once living forms or whether the resemblance is purely fortuitous. (From Ousdal.)

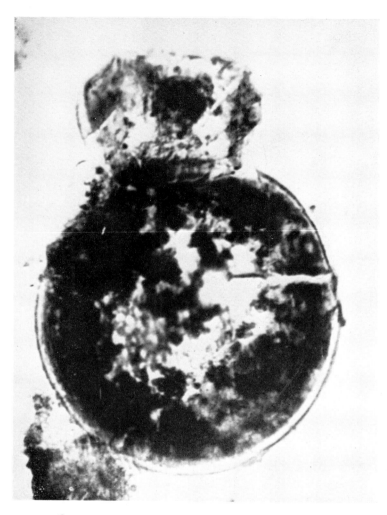

Fig. 1*b*. Photograph of material found in section of meteorite. Such sections may show shapes that resemble living forms. It is not clear if these in fact were once living forms or whether the resemblance is purely fortuitous. (From Ousdal.)

examined is equivalent to a small drop in a very vast ocean. It is pleasing that so much has already been discovered, but there is very little doubt that there is a great deal yet to be discovered about carbohydrate metabolism. It is therefore premature to claim that the "universal" occurrence of the glycolysis and citric cycles is proof of the common origin of life from one source.

To indicate some of the further biochemical complexities we may briefly mention four points. Firstly, it is often stated that all living systems use the same twenty or so amino-acids. This is a simplification of the known *data*. At one time it was thought that only the L-amino-acids occurred in natural systems, but since then a few D-amino-acids have been isolated. The number of known natural L-amino-acids has increased with the development of chromatographic techniques. Meister (1957) quotes some *seventy* naturally occurring amino-acids and he points out that new ones are being discovered almost every month! This is a result of the application of new techniques to an extended range of animal and plant material instead of restricting research to mammalian tissues.

Secondly, there are a large number of bacteria that use aberrant biochemical systems. Outstanding amongst these are the sulphur bacteria which grow quite well on water, carbon dioxide, phosphate and either sulphuretted hydrogen or sulphur. Another bacterium, *Thiobacillus ferro-oxidans*, can in some cases grow on ferrous iron under acid conditions which prevent the direct aerobic oxidation of ferrous iron. Other bacteria take ammonia and dehydrogenate it, or nitrite and oxidise it. There is some argument whether these systems are primitive or whether they are advanced and overlaid on the basic glycolysis and tricarboxylic cycles (see p. 22). These examples indicate that the metabolic systems in the bacteria are extremely varied.

Thirdly, even in the higher Metazoa the distribution of hydrogen acceptor systems such as in the cytochromes, flavoproteins, tocopherols, vitamin K, etc., is no more uniform than the distribution of blood pigments we mentioned previously.

A fourth generalisation that has been made about protoplasm is that its energetic systems involve the formation and destruction of "high energy" phosphorus compounds and the ubiquity of the phosphorus-containing compounds in living cells has been regarded as further evidence of a common protoplasmic origin. At first it was thought that the " high energy " compounds were found only in the form of ATP (adenosine triphosphate). But further studies have since shown the existence of many other " high energy " nucleoside triphosphates, e.g. guanosine triphosphate, cytidine triphosphate and uridine triphosphate. Recently other " high energy " compounds have been discovered which contain sulphur, i.e. acetyl coenzyme A. (Lynen 1952, Lipmann 1958). It is possible that further " high energy " compounds will be discovered in the future and this greater variety will make it less obvious that all protoplasm uses the same energetic systems.

Thus on the biochemical side it seems premature to conclude that all protoplasm has a common origin just because many cells show the components of the glycolysis cycle, citric cycle and the " high energy " phosphate compounds. It is likely that the protoplasm of different animals will show the presence of other schemes for the systematic degradation of carbohydrates and then perhaps in time an analysis of these systems will allow us to come to further conclusions about the varied metabolism of protoplasm.

Morphological evidence. A line of argument developed by morphologists to show the common origin of living cells is the almost universal occurrence of the mitotic and meiotic cycle. Thus Grassé (1952) suggests that such a system indicates the monophyletic origin of present-day animals and protozoa. But as Boyden (1953) pointed out, the mitotic cycle is not so fixed or so invariable as people imagine.

There are variations such as the presence or absence of intra- or extra-nuclear spindles and the presence or absence of centrioles. Thus Amano (1957) suggests that the chromosomes are separated by extending fibres in animal cells though a different mechanism exists in plant cells. On the other hand Swann (1951) suggests that the chromosomes in the *Arbacia* egg separate because of the contraction of fibres. In fact a perusal of Schrader's book *Mitosis* (1953) makes it quite clear that one difficulty in finding a single hypothesis to explain the mechanisms of mitosis in all cells is that there are a large number of different mechanisms of mitosis. It also seems that various tissues synthesise their DNA at different stages of the mitotic cycle and that the chromosomes may be duplicated at these various stages (Leuchtenberger 1958). It is possible that a more detailed examination of mitosis will show that it too is a polyphyletic system devised for the successful separation of the nuclear material into two equal sets. Whether one could go as far as Boyden (1953) and say " Under the circumstances the widespread occurrence of what is called mitosis or meiosis is no proof of real genetic relationships of all such organisms. On the contrary the very existence of such mechanisms in organisms otherwise so diverse as Protista, Metazoa, and Metaphyta, is strongly suggestive of convergence and may thus be interpreted with the theory of the strictly polyphyletic origins of the major groups of organisms " is another matter in our present state of ignorance.

What then can one conclude about the chemical and physical nature of protoplasm? Simply that we have a very great deal to learn about it. Modern developments are making it abundantly clear that some of our previous concepts are quite inadequate and that the picture is very much more complex than previously imagined. It would be a great mistake to assume that all is chaos and that there are no general common systems, but it would be a mistake of equal magnitude to assume that everything is very simple and that but one system will be found in all protoplasm. From our present viewpoint there would appear to be at least four or five different systems which allow a cell to obtain its energy. There are minor variations in this pattern and the higher animals may show less variation than do the bacteria (though few higher animals other than the pigeon and the rat have been studied). The picture in no way allows us to dogmatise and state that life in all its manifestations shows a common biochemical system indicative of a single genesis. The evidence at present does not by any means exclude the concept that present-day living things have many different origins.

Polyphyletic origin of life

If we do not hold that the origin of life was unique, i.e. life is monophyletic, there is the alternative point of view. This is that living things have been created many times, i.e. polyphyletic.

There are two ways of considering the multiple origin of life. The first is to consider that life is continuously being created all the time, i.e. that spontaneous generation is always occurring. The second view is that spontaneous generation occurred at some finite time in the past but that it is no longer occurring.

The continuous formation of life *de novo.* This theory is of considerable antiquity and it might be as well to give a brief *résumé* of its history. The responsibility for it is usually placed at the door of Aristotle. He wrote: " It is quite proved that certain fish come spontaneously into existence not being derived from eggs or copulation. Such fish as are neither oviparous nor viviparous arise all from one of two sources, from mud or from sand, and from decayed matter that rises hence as scum; for instance the so-called froth of small fry comes out of sandy ground. The fry is incapable of growing and of propagating its kind, after living for a while it dies away and another creature takes its place and so, with short interval excepted, it may be said to last the whole year throughout."

Other biologists gave various recipes for the formation of life *de novo.* Virgil in his *Georgics,* Book IV, gives the recipe for the formation of a swarm of bees from the barren carcass of a dead calf. Van Helmont suggested that mice could be formed " if a dirty undergarment is squeezed into the mouth of a vessel, within 21 days the ferment drained from the garment and transformed by the smell of the grain, envelops the wheat in its own skin and turns into mice." Van Helmont was surprised that mice formed in this manner could not be distinguished from mice produced by normal sexual breeding.

The situation became more critical when the experimentalists tried to determine whether it was possible to prevent living things from appearing in preserved material. The experiments of Needham, Pouchet and Bastian all indicated that living things still appeared in solutions from which all previous life had been removed, whilst Redi, Swammerdam, Vallisneri, Spallanzani, Schwann, Pasteur and many others showed that if the experiments were done very carefully it was possible to preserve soups, blood or urine in an atmosphere of oxygen and still get no growth of living material. It is not my intention here to discuss this old controversy. Full and interesting details can be found in the books of Oparin (1957), Singer (1950) and Wheeler (1939). Today there are still people who think that living things of a high level of complexity can be formed *de novo.* Of these it is perhaps of interest to quote from one Wilhelm Reich (1948). Reich has developed the concept that living material accumulates units of primordial energy which he calls " orgones." These orgones may be taken up by small vesicles (bions) that exhibit certain similarities to living material. By studying these bions and bion complexes under the very high optical magnification of 5,000 times ("it is not a matter of visualising finer structural detail but movement ") Reich concludes that bacteria and Protozoa can arise from sterilised organic and inorganic material. Thus from autoclaved grass he observed the development of amoebae and other Protozoa. Reich was not satisfied with the alternative explanation that the spores might have been present in the grass since he had also obtained similar amoebae from inorganic material such as sand or iron filings placed in the sterilised medium! The bions give off radiations which affect living material, and in some ways this radiation resembles the mitogenetic radiation studied by Gurwitsch (1926). It will be remembered that Gurwitsch claimed that the mitogenetic rays which come off from living cells affect the division rate of other cells. The experimental verification of mitogenetic radiation has proved to be very difficult and at best inconclusive; the evidence is summarised in Hollaender and Schoeffel (1931) and in Gray (1931), but as yet there has been no work on orgones other than from Reich and his colleagues. The work of Reich is of interest in that it shows that there are still " heretics " at work on the age-old problem of the origin and nature of living organisms.

Joseph Needham, writing in his textbook *Chemical Embryology* (1931) stated, " It may be remarked here, without irrelevance, that the problem (of spontaneous generation) is still unsolved; for all that was proved by the experiments of Spallanzani was that animals the size of rotifers and Protozoa do not originate spontaneously from broth, and all that was proved by those of Pasteur was that organisms the size of bacteria do not originate *de novo*. The knowledge which we have acquired in recent years of filter-passing organisms such as the mosaic disease of the tobacco plant, and phenomena such as the bacteriophage of Twort and d'Herelle has reopened the whole matter, so that of the region between, for example, the semi-living particles of the bacteriophage (10^{-15} g) and the larger-sized colloidal aggregates (10^{-18} g) we know absolutely nothing. The dogmatism with which the biologist of the early twentieth century asserted the statement *omne vivum ex vivo* was, therefore, like most dogmatisms, ill timed."

The argument developed so far, then, is as follows. The ancients thought that in many cases it was possible for living things to be created *de novo*. All these cases depended upon poor observation or lack of knowledge, and gradually as information has become available all the higher animals have been shown to arise from previous generations. The simpler forms of life such as yeast and bacteria were at one time thought to arise spontaneously, but controlled experiments showed that these observations were at fault. The conclusion is thus that the only cases where we think that life may be formed *de novo* are those where we have no information as to the mode of origin. From this one might suppose that spontaneous generation does not take place, but this is an unjustified extrapolation. The correct extrapolation would be that until we have devised experiments in which the simpler forms of life, such as viruses, are developed *de novo*, we have no evidence of *de novo* origin of life. This does not imply that *de novo* generation is or was impossible: Oparin (1957) suggests that life was created *de novo* on this world at one time and it is possibly being created now somewhere in the universe, but it is not being created now in this world since the ubiquitous presence of living bacteria would prevent the accumulation of the necessary raw materials for the formation of life *de novo*. When life was first created there were no such bacteria and hence the necessary substances accumulated. If we accept Oparin's view that life is not formed *de novo* at present in the world, there are still two alternative suggestions concerning the origin of life. The first is that life is still being formed *de novo* in other parts of the universe and is then transmitted by meteorites to this planet. Ousdal (1956) has described in meteorites some very interesting shapes which in some ways resemble present-day living forms (Fig. 1). However, the meagre evidence so far available that meteorites may contain living material is not yet convincing. It should be noted that the present climate of opinion concerning the possible mechanism of the evolution of the present solar system is changing. The view suggested by Sir James Jeans that the planets were formed by the unique passage of a giant star near to the sun is no longer strongly supported (Lyttelton 1956). Instead it seems that solar systems similar to our own have been created many millions of times and thus conditions favourable to life may be present on many other planets in the universe. Shapley (1957) has calculated that there are probably 10^8 planets that have conditions favourable for life of one sort or another. We have no evidence that living things can be transmitted between the stars and still remain alive on reaching their destination, but it would seem that we shall shortly have information on transmission of living material by rockets. It would perhaps be more to the point to have information concerning transmission by meteorites. Oparin (1957) gives quite an extended discussion of meteorite transmission of life and concludes that it was, and still is, highly improbable. Nevertheless, at present we have very little information on this subject and it is likely that the renewed interest in space travel will stimulate further investigation into the nature and properties of meteorites.

Unique occurrence of life. The second suggestion is that though we are unable to show at present that life is formed *de novo* on this earth, there is no evidence to show that when life was formed on this earth it was a unique event. Haldane (1954) and Oparin (1957) are of the opinion that life was uniquely formed, but, as they both point out, nothing is definitely known about what did happen; all is hypothesis, and though it is simpler to assume that it was a unique occurrence there is no reason why this simple explanation should be the correct one. In the previous pages it has been pointed out that our knowledge of the cell metabolism is insufficient to allow us to state categorically that all cells in all living forms have the same biochemical systems at work. Though the similarities are often great, the dissimilarities may be just as impressive.

If living material had developed on several different occasions or at different places at the same time, then one would expect to have a large number of distinct groups of animals, whose relationships and affinities are difficult to determine.

This, as we shall see, is the present situation.

CHAPTER 3

VIRUSES, RICKETTSIAE AND BACTERIA

IF ONE assumes that the origin of life was a unique occurrence then it follows that all the present-day living things must be derived from this original source. This then poses the problem, " What is the relationship between the present-day forms? " In many cases it is difficult to form any definite conclusion regarding these relationships and this certainly seems to hold for the relationship between Viruses, Rickettsiae and Bacteria.

The viruses

The viruses are of interest since they show many of the properties of living material. At first they were described as material that would pass through a bacterial filter and which was capable of reproducing in the living cell. But later on considerable confusion arose over the chemical nature of viruses, the main trouble being one of over-simplification. Many people thought that the viruses were necessarily simple because they had been prepared in a crystalline condition. This concept was furthered when chemical analysis showed that the virus was composed of a " simple chemical substance "—nucleoprotein. With more advanced techniques it became clear that there was considerable variability in virus structure. Markham, Smith and Lea (1942) showed that when the tobacco mosaic virus was irradiated, only a small part of the virus proved sensitive to radiation. This part was some 5%–6% of the virus area and in effect it behaved like the nucleus of the virus. In 1951 Markham and Smith presented evidence that the turnip mosaic virus contained at least two distinct components, a nucleic acid component (38%) and a structural protein component (62%).

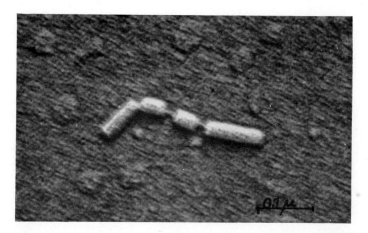

FIG. 2. Virus structure. The tobacco mosaic virus is made up from at least two components. There is a central rod of nucleic acid and a series of units of protein that fit over the central rod. In the photograph shown here part of the second component has been dissolved away to reveal the nucleic acid. (This photograph was obtained through the kindness of Professor G. Schramm.)

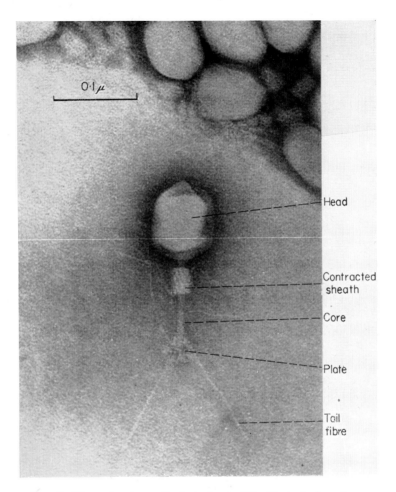

0·1μ

Head

Contracted
sheath

Core

Plate

Tail
fibre

FIG. 3. Virus structure—Bacteriophage. The Bacteriophage has a
more complex structure than a simple virus such as the tobacco
mosaic virus. It has a well-developed head and a tail. The head
contains the nucleic-acid component and this flows through the
tail into the bacterium it attacks and there reproduces. (This
photograph was obtained through the kindness of Dr. S. Brenner
and Dr. R. W. Horne.)

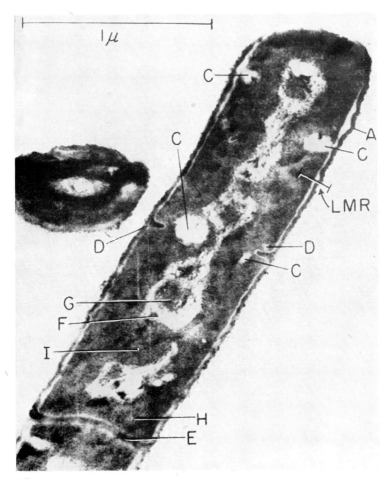

Fig. 4. Bacterial structure. The bacterial cells have complex
structure as is shown by this electronmicrograph of a section
through *B. cereus*. The cells have a well-defined cell wall, a
nuclear structure, and many types of cell inclusions. (From
Chapman and Hillier.)

(A) Cell wall.   (F) Fibrous material.
(C) Peripheral bodies.   (G) Nucleus.
(D) Transverse cell wall.   (H) Cytoplasm.
(E) Transverse cell wall.   (I) Inclusions.

L.M.R. Limit of light microscope.

Takahashi and Ishii (1953) showed that it was possible to find the structural protein in the sap of the plants infected with the mosaic virus and that it differed from the normal plant proteins. This protein had no power of reproduction but required the presence of nucleic acid. If the nucleic acid was added to the structural protein, then it became capable of reproduction inside the cell (Fraenkel-Conrat and Williams 1955).

The chemical analyses of virus structure have been paralleled by studies using the electron microscope. These show that the tobacco mosaic virus is often found in rod-like forms, the rods being made up of a series of discs each with a hole in the centre. The hole is apparently filled with the nucleic acid whilst the disc itself is probably the structural protein (Fig. 2).

Other viruses such as bacteriophage which attacks bacteria have an even more complex structure. The bacteriophage has a tadpole-shaped head and a small tail (Fig. 3). The head consists of a shell of structural protein inside which is the nucleic acid. Hershey (1956) described how it was possible to remove the nucleic acid from the bacteriophage and leave the tail and the shell. This skeleton was still capable of attacking a bacterium and killing it, but it was not capable of self-reproduction.

Detailed chemical analysis and electron microscope studies have therefore shown that viruses are not simple single chemical substances. There is a considerable range of structural and chemical complexity within the group of viruses and it is possible to draw up a table showing the differences in their chemical composition.

Material Present   Virus

RNA
DNA
Protein   } Animal virus
Fats
Carbohydrates

RNA
DNA   ?
Protein   } Bacteriophage
Fats

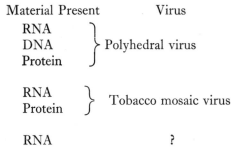

Material Present   Virus

RNA ⎫
DNA ⎬ Polyhedral virus
Protein ⎭

RNA ⎫
Protein ⎬ Tobacco mosaic virus

RNA   ?

Thus the analysis of the vaccinia virus shows the presence of proteins, DNA, neutral fat, phospholipid, cholesterol, biotin, flavine, copper and various as yet unidentified substances. The Lee influenza virus has about 5% of its weight as a complex polysaccharide containing mannose, galactose and glucosamine. It is also becoming clear that the term " nucleic acid " should be used with care since there are many different nucleic acids, and as Chargaff (1957) points out, often the term " ribose nucleic acid " is used when there is no evidence that the sugar ribose is present. The precise structure of the nucleoproteins is not yet known, i.e. the type of proteins, and the way in which the nucleic acid is attached to the protein have yet to be fully elucidated, but some evidence is available concerning the component nucleotides in the nucleic acids of the virus. The table below, taken from Fruton and Simmons (1958) indicates that the proportional composition of the nucleotides varies in the different viruses.

Virus	Molar proportions in nucleic acid			
	Adenylic acid	Guanylic acid	Cytidylic acid	Uridylic acid
Tobacco mosaic virus	1·0	0·89	0·65	0·88
Cucumber mosaic virus	1·0	1·0	0·75	1·15
Tomato bushy stunt virus	1·0	1·0	0·74	0·89
Turnip yellow mosaic virus	1·0	0·76	1·68	0·98

This gives some hint of the complexity of nucleic acid, and nucleoprotein structure, and makes one careful when ascribing simplicity to a system that is not yet adequately understood.

There are two main views concerning the nature of viruses. One suggests that they are in fact the simplest and most primitive forms of living material and that originally they utilised the proteins found in the complex primaeval " soup." As they gave rise to more complex living things which altered and destroyed the primaeval soup, so they became obligate parasites in other living systems that evolved along different lines. On the other hand there is the view that the viruses arose from more complex systems and that in effect they are more like genes that have taken on a free-lance life. Both of these views are discussed by Luria (1953).

There are other opinions concerning the nature of viruses. Thus Hadži (1953), for example, has suggested that viruses are the spores of parasitic Protozoa. It is possible that *all* these opinions are correct and that the viruses are a complex group of substances at present classified by their properties and that these properties depend on the level of organisation that has been achieved. The viruses are thus most likely a grade of organisation that has been reached from many different directions.

In this context and throughout the book, a *grade* may be regarded as a group of individuals that are united by certain common properties but are not derived from a common close ancestor. The grade indicates the level of organisation rather than a close phylogenetic relationship.

The rickettsiae

The rickettsiae cause such diseases as typhus, murine fever and spotted fever. They have properties between those of bacteria and viruses; they approach the bacteria in structural complexity and size, and they resemble viruses in that they are unable to reproduce outside living cells (though this is not a stringent criterion; it merely indicates lack of experimental success so far).

The rickettsiae are more complex than viruses in that they are able to carry out certain of the metabolic processes of the higher cells. Thus they are capable of oxidising glutamate, pyruvate, succinate, fumarate and oxalo-acetate. These substances are also oxidised by the mitochondria of the normal cell, and the suggestion has been made that the rickettsiae are in fact free mitochondria. Thus both the mitochondria and rickettsiae lose their diphosphopyridine nucleotide and coenzyme A on freezing, the freezing in some way affecting the properties of the membrane around the rickettsiae or mitochondria. It is possible that the rickettsiae are developed as free mitochondria and that the viruses are further simplifications. On the other hand the rickettsiae may indicate a stage in the development of the viruses to bacteria or the three groups could be quite unrelated.

We have insufficient evidence as yet to come to any firm conclusion concerning the origin and affinity of the rickettsiae.

Bacteria

We are no wiser when we come to consider the status of the bacteria. Within recent years there has been a considerable increase in our knowledge of the structure of the bacterial cell (Spooner and Stocker 1956; Zinsser 1957). Thus Robinow in 1946 suggested that there were certain components within the cells of *Escherichia coli* that behaved like nuclear material during cell division (Fig. 4). Lederberg (1947) showed that a type of crossing over occurred between certain strains of *E. coli* and that in effect it was possible to draw up a map of the positions of various factors in bacterial metabolism. The conclusion, then, is that certain bacteria show nuclear and sexual (parasexual) behaviour. On the other hand there are many bacteria that do not show these phenomena, their structure and life history being much more simple.

It is not clear whether the bacteria represent an evolutionary approach to the Protozoa, whether they are a retreat from the Protozoa or whether they are quite unrelated. Perhaps some of the difficulties can be illustrated by considering the autotrophic bacteria (Chemoautotrophic) (Fry and Peel 1954). These bacteria such as the sulphur and iron bacteria are able to metabolise various simple substrates. They raise the question " are these bacteria using a more primitive (earlier developed) system than those found in the heterotrophic and photosynthetic bacteria? " It is not possible to give a definite answer to this question since our knowledge of the biochemistry of the heterotrophic and chemoautotrophic bacteria is still very incomplete. The chemoautotrophs can obtain their energy from simple sources such as hydrogen, methane, ammonia, nitrite, hydrogen sulphide or iron compounds. These substances are very much less complex than the carbohydrates from which the higher animals obtain their energy. The simple hypothesis is that the chemoautotrophs are a side-line representing a more primitive state of development than that shown by heterotrophic and photosynthetic bacteria. Though this opinion is quite widely held, evidence is gradually accumulating to indicate the opposite view; viz. that the chemoautotrophs are in fact using systems that are secondarily simplified from those of the heterotrophs. Thus O'Kane (1941) showed that the sulphur bacterium *Thiobacillus thioxidans* could synthesise various vitamins of the B group. These substances are used mainly in normal heterocyclic heterotrophic glycolysis; thiamine is used in oxidative decarboxylation; riboflavine is a coenzyme for the hydrogen acceptors, nicotinic acid forms part of Coenzymes I and II. It would therefore be interesting to know what role they play in *Thiobacillus*. It would appear that the bacterium has many of the enzymes that are used in heterotrophic glycolysis but that it uses special variations on the normal system. The chemoautotrophs would then have superimposed their own system upon that of the heterotrophs.

A schematic system for the development of metabolic systems is shown below. If this is correct, and the chemoautotrophs are less primitive than the heterotrophs, it again points the lesson that the simplest explanations are not necessarily the correct ones.

Scheme for the origin of metabolic systems (after Oparin):

(1) Solution containing salts.
(2) Solution containing salts and simple organic compounds.
(3) Solution containing salts, simple and complex organic compounds.
(4) System that turns complex materials into simple organic materials and so obtains energy. Also able to reproduce itself $=$ a living system.

> HETEROTROPHS (only glycolysis cycle)

(5a) Living system that converts complex organic material to simple material.

> HETEROTROPHS (glycolysis and citric cycles. Hydrogen acceptors)

(5b) Living system that converts simple material to obtain energy.

CHEMOAUTOTROPHS (attack H_2S, CH_4, etc.)

(5c) Living system that develops PHOTOSYNTHESIS. (Note that animals can by chemical means build up CO_2 to form carbohydrates.)

Bacteria and Protozoa

It is problematical how, if at all, the Bacteria are related to the Protozoa and, if so, which Bacteria gave rise to which Protozoa.

Grassé (1953) thinks that the Protozoa are in fact monophyletic and derived from the Bacteria. He bases this opinion on the following resemblances between the Protozoa and Bacteria.

(1) Both have vacuoles.
(2) Both contain proteins, lipids and carbohydrates.
(3) Both have mitochondria.
(4) Certain bacteria have a nucleus and chromosomes.
(5) A sexual process has been described in some bacteria.
(6) Both can possess flagella.
(7) Spore formation occurs in both.
(8) The membranes around the cell in each case are sometimes morphologically similar.

These resemblances are rather tenuous and not all apply to any one bacterium. In effect it is difficult to know to what extent the resemblances are real phylogenetic ones and to what extent they have risen by convergence. Thus the bacterial flagellum is very much more simple in structure than the protozoan flagellum. The protozoan flagellum has an inner strand of two rods and an outer ring of nine rods. The bacterial flagellum has just the inner strand of one or two rods. Until we know a great deal more about the electron microscopy of the bacterial and protozoan cells we shall not be in any position to base relationships on morphological similarities.

The relationship between Bacteria, Protozoa and Metazoa has been discussed by Grassé, who derives the following alternative systems:

Autotrophic bacteria → Protophyta →Metaphyta

(1)   ↓   ↓

Heterotrophic bacteria   Protozoa   → Metazoa

(2) Autotrophic bacteria   → Protophyta → Metaphyta

Heterotrophic bacteria → Protozoa   → Metazoa

Autotrophic bacteria   Protophyta → Metaphyta

(3)   ↓   ↑

Heterotrophic bacteria → Protozoa   → Metazoa

Autotrophic bacteria   → Protophyta → Metaphyta

(4)   ↓

Heterotrophic bacteria → Protozoa   → Metazoa

Grassé suggests that the first system is the most probable but that all the others have something to be said in their favour. On the other hand Oparin (1957) thinks that the heterotrophic bacteria are the most primitive and that they gave rise to the autotrophic forms and to the Protozoa and Protophyta, a view that is not mentioned in Grassé's scheme. It can be seen that nothing is definite; all is hypothesis and opinion.

At present the following schemes all seem equally likely:

(1) The Bacteria and the Protozoa had an independent phylogenetic origin.
(2) The Bacteria are more primitive than the Protozoa and gave rise to the Protozoa.
(3) The Bacteria are secondarily simplified and were derived from the Protozoa.
(4) The Bacteria are a polyphyletic grade. Some are more primitive than the Protozoa, others are derived from the Protozoa.

We have at present insufficient evidence to enable us to choose between these hypotheses.

CHAPTER 4

THE PROTOZOA

WE HAVE just seen that the relationship between the simplest living forms, the Viruses, Rickettsiae and Bacteria, is not at all clear. We cannot say with any certainty how they have evolved and what the relationship is between the three groups. When we come to consider the next group of animals, the Protozoa, we shall find a very similar situation. There is great difficulty in deciding where the Protozoa came from, what they gave rise to, and what their interrelationships are. The Protozoa can be classified into four classes. These are:

(1) Flagellata; e.g. *Chlamydomonas, Trichonympha.*
(2) Rhizopoda; e.g. *Amoeba, Elphidium.*
(3) Sporozoa; e.g. *Monocystis, Plasmodium.*
(4) Ciliophora; e.g. *Paramecium, Entodinium.*

There are several evolutionary problems to be found in the protozoans but only three of these will be considered here. The first problem is, " Which of the four classes is the most primitive? " The second problem is, " What is the interrelationship of the four classes? " and the third problem is, " What is the status of the group Protozoa? " These problems will each be examined in turn.

THE MOST PRIMITIVE PROTOZOA

We can readily dismiss two of the four classes of the Protozoa as candidates for the position of the most primitive class. The Sporozoa are almost entirely parasitic in the bodies of higher animals and they spend their life in the outside world in the encysted state. Though it is possible that they could have arisen in the " primaeval soup " that gave rise to living forms, their life cycles are so involved and their structure with myonemes and spore cases so complex that they are probably not very close to the primitive stock. It is worth noting here that some authors such as Ulrich (1950) have suggested that the Cnidosporidia are not protozoans but metazoans.

The Ciliophora too can be disregarded since they show *par excellence* the extremely complex structures that can exist within the protozoan cell. Thus *Entodinium* with its complex cirri, neuromotor system, skeleton, nuclei and digestive system is almost as complex as some metazoans (Fig. 5). On the other hand even the most simple of the ciliophorans have a complex nuclear structure. There is usually a macro- and a micro-nucleus as separate bodies, though they are in the form of macro- and micro-chromosomes in a single nucleus of the Chonotricha such as *Spirochona*. In addition there is the very complex infraciliature that has developed in the superficial regions of the ciliates and this is more complex than that found in the flagellates.

This then leaves two classes, the Flagellata and the Rhizopoda, as the more primitive protozoans and each of these has at various times been considered as the most primitive Protozoa. Thus at the beginning of the century the prevalent view was that the Rhizopoda were the most primitive of the Protozoa. This view was well expressed by Ray Lankester (1890) in his article in the *Encyclopaedia Britannica*. Lankester said, " Briefly stated the present writer's view is that the earliest protoplasm did not possess chlorophyll and therefore did not possess the power of feeding on carbonic acid. A conceivable state of things is that a vast amount of albuminoids and other such compounds had been brought into existence by those processes which culminated in the development of the first protoplasm, and it seems therefore likely enough that the first protoplasm fed on these antecedent steps in its own evolution just as animals feed on organic compounds at the present day, more especially as the large creeping plasmodia of some Mycetozoa feed on vegetable refuse. It is indeed not improbable that, apart from their elaborate fructification, the Mycetozoa represent more closely than any other living forms the original ancestors of the whole organic world. At a subsequent stage in the history of this archaic living matter chlorophyll was evolved and the power of taking carbon from carbonic acid. The

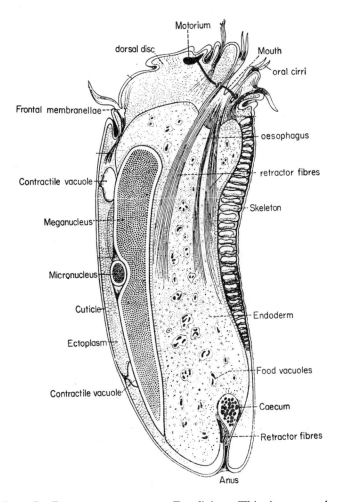

FIG. 5. Protozoan structure.—*Entodinium*. This is a complex ciliate and has much of the differentiation that one expects to find in the higher animals. Thus it has a mouth, gullet, cloaca, myonemes, neuronemes, contractile vacuoles, skeletal system and several nuclei. (After C. V. Sharp.)

' green ' plants were rendered possible by the evolution of chlorophyll, but through what ancestral forms they took their origin or whether more than once, i.e. by more than one branch, it is difficult even to guess. The green Flagellate Protozoa (Volvocinae) certainly furnish a connecting point by which it is possible to link on the pedigree of green plants to the primitive protoplasm; it is noteworthy that they cannot be considered as very primitive and are indeed highly specialised forms as compared with the naked protoplasm of the Mycetozoön's plasmodium. Thus we are led to entertain the paradox that though the animal is dependent on the plant for its food yet the animal preceded the plant in evolution, and we look among the lower Protozoa and not among the lower Protophyta for the nearest representatives of that first protoplasm which was the result of a long and gradual evolution of chemical structure and the starting point of the development of organic form."

If one consults any of the older texts such as those of Lankester (1909), Delage and Hérouard (1896) or Kukenthal and Krumbach (1923) one finds that the Rhizopoda are placed as the first class of the Protozoa.

The accent of protozoan research changed during the first part of the twentieth century. Instead of being concerned with the morphology and life cycles of the Protozoa, the interest became more centred upon the physiology and in particular the nutritional requirements of the Protozoa. This change in accent from a morphological one to a physiological one may explain the change that took place in the prevalent attitude to the phylogeny of the Protozoa. In such texts as those of Hyman (1940) or Grassé (1952) the Flagellata take pride of place over the Rhizopoda; the Flagellata being the first class to be described. It should, however, be noted that Klebs in 1892 suggested that the Flagellata were in fact more primitive than the Rhizopoda.

Grassé points out that since many of the members of the Flagellata possess chlorophyll they are able to undertake synthesis of all their food requirements without the assistance of any complex compounds. This view is much the same as that of Pringsheim (1948), who showed that many of the colourless flagellates such as *Astasia* or *Polytoma* can be found in pure cultures of *Euglena* and *Chlamydomonas* respectively. The coloured forms gave rise to the colourless forms, i.e. *Astasia* is a colourless *Euglena*, and hence the groups *Astasia* and *Polytoma* are not strict monophyletic genera but instead are polyphyletic grades. Pringsheim suggests that many of the present-day colourless flagellates are derived from the coloured form and this would make the coloured forms more primitive than the colourless forms (Pringsheim and Hovasse 1950).

Lwoff (1944) in his book on physiological evolution goes even further and contends that from a physiological point of view evolution is retrogressive. The most primitive Protozoa, he states, must surely have been entirely self-supporting with little or no food requirements, but as evolution occurred the cells lost their synthetic ability and became more and more dependent upon other cells for the provision of their food requirements; i.e. they regressed instead of progressed.

There appears to be a fallacy in Lwoff's argument. The fact that a cell has minimal food requirements does not mean that this is necessarily the most primitive condition. In fact most of the schemes suggested for the origin of living material place the advent of chlorophyll at a very late stage in the evolutionary sequence, the plant cells having the chlorophyll system superimposed on the anaerobic metabolic system (Oparin 1957). The very earliest living forms would have had considerable food requirements. It would be perfectly possible for a sarcodine-like form to be the most primitive animal feeding on amino-acids and carbohydrates synthesised by abiogenic methods. The presence of chlorophyll is indeed a good reason for considering the Flagellata as an advanced group of the Protozoa. This view was in fact suggested by Lankester in 1909.

Lankester stated, " The real question . . . is whether we find reason to suppose that the combination of carbon and nitrogen to build up proteid, and so protoplasm, required in the earliest state of the earth's surface, the action of sunlight and the chlorophyll screen. We must remember that these are now necessary for the purpose of raising carbon, and indirectly nitrogen, from the mineral resting state to the high elaboration of the organic molecule, yet it is, after all, living protoplasm which effects this marvel with their assistance; and it seems (though possibly there are some who would deny this) that it is protoplasm which has, so as to speak, invented or produced chlorophyll. Accordingly I incline to the view that chlorophyll as we now know it is a definitely later evolution—an apparatus to which protoplasm attained, and as a consequence of that attainment we have the arborescent, filamentous, foliaceous, fixed series of living things we call plants. But before protoplasm possessed chlorophyll it had a history. It had in the course of that history to develop the nucleus with its complex mechanism of chromosomes, and it had during that period to feed."

There is a second reason why the Flagellata are sometimes considered to be more primitive than the Rhizopoda. During their young stage some of the Rhizopoda such as *Naegleria* and *Dimorpha* show a flagellate condition which is considered by some investigators to be a form of recapitulation; i.e. the young stage shows more primitive characteristics than those present in the adult. How much faith can one have in this type of argument? We know that certain flagellates such as *Mastigamoeba* show pseudopodia as well as flagella and thus the presence or absence of pseudopodia or flagella does not necessarily indicate primitiveness. What is more important is the concept that these Rhizopoda show the flagellate stage only in the young forms and that therefore the flagella are more primitive than the pseudopodia.

We are lucky in that there has been a recent investigation by Willmer (1956, 1958) into the factors that determine the acquisition of flagella by the amoeba *Naegleria gruberi*. Willmer showed that the amoeboid *Naegleria* can be made to turn into a flagellate form with one to four flagella by placing it in water. The change takes from 20 min to 24 hr to complete and during this time is accompanied by the development of a definite antero–posterior axis in the cell; the flagella appearing at the anterior end. The pseudopodia can develop from any part of the animal. The presence of salts such as lithium chloride, magnesium chloride and magnesium sulphate suppress the development of the flagella but leave the pseudopodia fully active. The change from amoeboid to flagellate condition is reversible and depends upon the environmental conditions (Fig. 6). This means that the flagellate condition is not necessarily found in the young animal; either stage can reproduce and either stage can be found in the young animal. Bunting (1926) showed that the rhizopod *Tetramitus* could undergo cell division in either the amoeboid or the flagellate

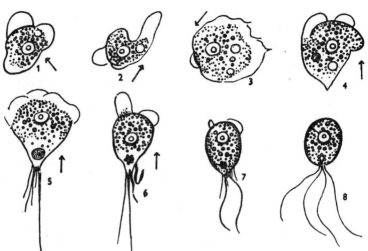

FIG. 6. *Naegleria gruberi*. This protozoan can exist in either of two forms: an amoeboid form or a flagellate form. Stages 1–8 show stages during which the amoeboid form changes into the flagellate condition. The arrow indicates the direction in which the animal moves. (From Willmer.)

stage. This too indicates that the flagellate stage is not necessarily the more juvenile one.

On the basis of this evidence we are left undecided as to which is the most primitive, the Flagellata or the Rhizopoda. This question will be dealt with again on p. 33, where a third interpretation will be presented. In effect this third view states that the Rhizopoda and Flagellata are not strict classes of the Phylum Protozoa. Instead they are polyphyletic grades. The Flagellata arose on many separate occasions from the plants, fungi and metazoa, and the Rhizopoda developed in much the same manner. Both these groups are then more in the nature of horizontal grades than vertical monophyletic classes, one of which is older than the other.

PROTOZOAN PHYLOGENY; THE INTERRELATIONSHIP OF THE FOUR CLASSES OF PROTOZOA

The precise relationship of the four classes of Protozoa is uncertain. The two classes that appear to be the most closely related are the Flagellata and the Rhizopoda. Bütschli in 1883 suggested that it was possible to derive these two classes from intermediate forms such as *Mastigamoeba*, and this view has been followed by Grassé in his *Traité de Zoologie*, where he groups the Flagellata and the Rhizopoda into a subphylum: the Rhizoflagellata. To the groups Flagellata and Rhizopoda he gives superclass status and groups such as the Dinoflagellata and the Foraminifera are termed Classes.

The Sporozoa were linked by such workers as Doflein (1916) with the Flagellata and the Rhizopoda to form the group Plasmodroma. There are certain resemblances between these groups. Sporozoans such as *Plasmodium* have both flagellate sperm and amoeboid oökinetes, and spore formation is found in both the Flagellata and the Rhizopoda. The Plasmodroma are then separated from the Ciliophora with their complex infraciliature. Yet even within the Ciliophora there are forms that are possibly related to or have something in common with the Flagellata. Thus *Opalina* is according to some writers a ciliate and according to others such as Grassé it is a flagellate.

Such close connexions between the four classes can be interpreted as showing how closely the various groups are related. But there is another interpretation. Franz (1924) has suggested that the Protozoa are not a strict phylum but instead are a grade of organisation. He thinks that there is no good evidence that the Protozoa are more primitive than the Metazoa and states that the unicellular forms could have been derived many times from the Fungi, Algae and the Metazoa. The various groups such as the Flagellata, Rhizopoda, Sporozoa or Cilophora would then each be polyphyletic and contain animals that have been derived from different sources at different times but which are grouped together because they have certain convergent morphological characteristics.

The view that the four classes are polyphyletic is discussed by Hyman (1940). " The flagellates themselves appear to be a heterogeneous assembly of groups that have probably arisen from a number of different sources, possibly bacteria and spirochaetes, many of which are provided with flagella. . . . The rhizopods like the flagellates constitute an arbitrary assemblage of forms having in common the pseudopodial method of locomotion and food capture. It is probable that the various orders of rhizopods have arisen independently from the different groups of flagellates, i.e. the class is

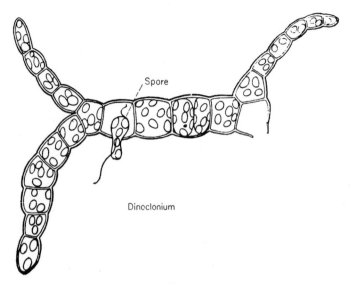

Spore

Dinoclonium

(A)

(B). Gymnodinium

Fig. 7. Relationship between the Protozoa and Algae. The alga
Dinoclonium has a flagellate spore that resembles a dinoflagellate
such as *Gymnodinium*. It is suggested by some authors that
Gymnodinium is more closely related to *Dinoclonium* than it is to,
say, *Amoeba*. (From Grassé.)

polyphyletic. . . . The Sporozoa are again a heterogeneous group of which the different orders have probably had separate origins. . . . The Ciliata differ so markedly from the other Protozoa in their possession of cilia, nuclear dimorphism, and sexual phenomena that their relation to them remains problematical."

So of the four classes of the Protozoa we see that at least three are suggested by Hyman as being polyphyletic.

Baker (1948) has similar doubts about the status of the Protozoa. In particular he considers the relationship of the dinoflagellate *Gymnodinium* with the filamentous alga *Dinoclonium* (Fig. 7). During the life cycle of *Dinoclonium* it develops spores almost indistinguishable in structure from *Gymnodinium*, but *Dinoclonium* is placed in the Algae whilst *Gymnodinium* and *Amoeba* are placed in the Protozoa. The structure of these spores clearly shows that *Gymnodinium* is more closely related to *Dinoclonium* than it is to *Amoeba*. Baker concludes that the Protozoa cannot be a mono-phyletic group.

From the evolutionary point of view we therefore have several problems in the Protozoa.

(1) The Protozoa do not seem to be a group of closely related animals. It is most likely that they are a polyphyletic group and the name " Protozoa " indicates a grade or status rather than a natural taxonomic group. In this they would be analogous to the group " Vermes " or " Pisces "; i.e. they show a level of organisa-tion and not an evolutionary relationship. (We shall see that this problem arises again and again; many of our phyla and classes are grades of animals that are not closely related.)

(2) It is difficult to decide which of the Protozoa are the most primitive. The information at our disposal is not sufficient to allow us to come to any definite conclusion.

(3) Each of the four classes probably contains the results of convergent development from heterogeneous stocks.

CHAPTER 5

ORIGIN OF THE METAZOA

WHEN the basic assumptions underlying evolution were discussed on p. 13 it was pointed out that if the modern living forms were polyphyletic, it should prove difficult to decide their interrelationships and we should have a number of isolated groups of animals. This is precisely what we have discovered so far. The Viruses, Rickettsiae, Bacteria and Protozoa are all quite distinct from one another and their interrelationship is anything but clear and certain. We come now to the Metazoa and we have to decide whether they can be linked to any of the lower groups of animals.

There are three main views concerning the origin of the Metazoa. These are that the Metazoa arose from (1) the colonial protozoans, (2) the syncytial protozoans, and (3) the Metaphyta.

Let us consider each of these views in turn.

(1) Origin from colonial Protozoa

Though the Protozoa are often defined as unicellular animals there are many protozoans which after division or budding do not separate their progeny so that the adult develops a colonial or multicellular form. This development into colonies has taken place many times within the Protozoa, as can be seen by looking at the various classes. The most common examples are found in the Flagellata. Simple unicellular forms such as *Chlamydomonas* can at times exhibit an aggregated stage. An example of this is the palmella stage during which *Chlamydomonas* encysts and divides asexually; the results enclosed in a gelatinous case may be regarded as a colonial form (Fig. 8).

In *Gonium sociale* a group of sixteen *Chlamydomonas* like individuals are associated together in a plate. All these forms are

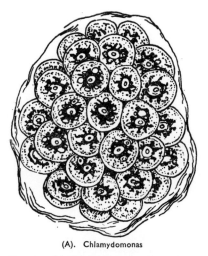

(A). Chlamydomonas

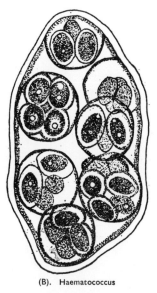

(B). Haematococcus

FIG. 8. (A) Palmella stage of *Chlamydomonas*. During this stage the protozoan divides but the cells remain together enclosed in a gelatinous case. The stage is " multicellular " though it is in fact a resting stage in the life history of the protozoan. (From Grassé after Goroshankin.)

FIG. 8. (B) Palmella stage of *Haematococcus*. (From Grassé after Wollenweber.)

alike and at reproduction each of them divides and forms gametes. A more complex colony is that of *Eudorina* in which there are sixty-four individuals (Fig. 9).

Pleodorina illinoiensis and *Pleodorina californica* show further stages in the development of the colony in that a group of cells become differentiated from the others and they are unable to take part in reproductive activities. There are four somatic cells in *Pleodorina illinoiensis* and thirty-two in *Pleodorina californica*. The soma is even more developed in *Volvox*, where the majority of the cells are unable to take part in the reproductive activity. The cells beat their flagella in a co-ordinated manner so that the colony can be regarded as having an antero–posterior axis and a dorso–ventral axis. The dorsal cells are slightly larger than the ventral cells and the colony during locomotion moves slowly through the water and does not turn over and over (Fig. 10).

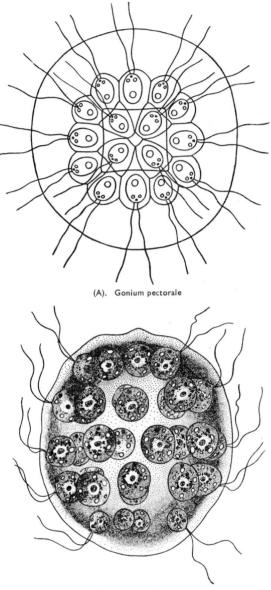

(A). Gonium pectorale

(B). Eudorina illinoiensis

FIG. 9. Colonial flagellates. Flagellates such as *Gonium* and *Eudorina* exist in a colonial form. The colony is active and thus differs from the palmella stage shown in Fig. 8.

(A) From Grassé after Migula. (B) From Grassé after Merton.

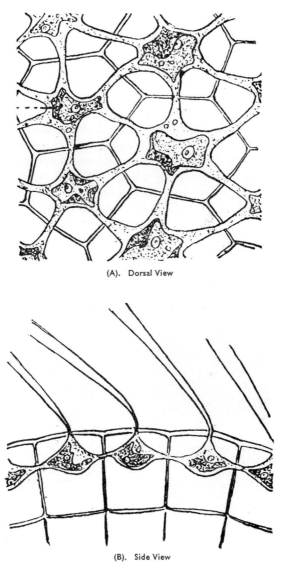

(A). Dorsal View

(B). Side View

FIG. 10. Colonial flagellates. *Volvox* though a colonial animal has protoplasmic connexions between the units of the colony. In this respect it can be regarded as a syncytium. (From Borradaile and Potts, after Janet.)

These examples are merely illustrative phases of the development of the colonial habit. There is no evidence that *Eudorina* gave rise to *Pleodorina* or *Volvox*.

Further examples of the development of colonial stages are found in the dinoflagellates. Though some forms such as *Gymnodinium* are solitary, others such as *Ceratium* at times may form long chains of individuals joined together in a temporary manner. *Polykrikos* is of interest since it shows a more permanent attachment. *Polykrikos schwartzi* usually contains four nuclei and a series of associated sets of flagella. It possesses cnidocysts which are manufactured within the cell and are used in catching prey. There is also a cytoplasmic connexion between the units of the *Polykrikos* colony. Thus in the related genus *Pheopolykrikos beauchampi* when the animal is touched it contracts up and clearly shows that there are four units in interconnexion (Fig. 11).

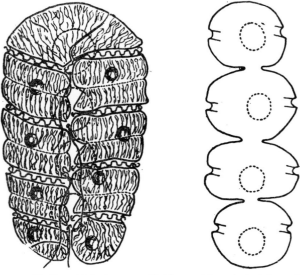

(A). Normal Animal   (B). Diagram of shape of animal after tactile stimulation

FIG. 11. Colonial dinoflagellates. The dinoflagellate *Pheopolykrikos* is clearly made from four dinoflagellate units. When it is touched it changes its form (B) and the four units can be distinguished. (From Grassé after Chatton.)

Another interesting colonial dinoflagellate is the parasitic form *Haplozoön*. This is found in the gut of polychaetes. It forms first of all a small cell which attaches to the gut of the host by means of a spike and some filamentous pseudopodia. This cell absorbs food from the polychaete and at a later stage divides. The results of division do not detach but instead remain in contact so that a colonial form of up to several hundred cells is soon formed, the number of cells differing from species to species. These cells can form a three-dimensional mass with small spaces between the cells through which food particles can be transferred (Fig. 12).

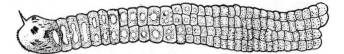

Fig. 12. Colonial dinoflagellates. *Haplozoön* was originally placed in the Mesozoa but its spores have typical dinoflagellate structure and it is now considered to be a colonial dinoflagellate. (From Grassé after Dogiel.)

The relationship of *Haplozoön* to the Dinoflagellata is not clear at first sight. In fact *Haplozoön* is so much like a metazoan that when it was discovered by Dogiel in 1906 he placed it in the Catenata, a new group of the Metazoa. It was not until the work of Chatton (1920) that it was shown that the cells at the posterior end of *Haplozoön* detached and developed into four small spores, each of which had characteristic dinoflagellate structure. It is for this reason that *Haplozoön* is placed in the Dinoflagellata. This reasoning can, if carried to its illogical conclusion, lead one into difficulties. Thus Duboscq and Grassé (1933) have shown that the mammalian spermatozoan is very much like a protozoan of the group Bodoines, yet I doubt if anyone would like to place Man in the Protozoa on account of his male gamete!

Colonial Protozoa are also found in the other classes. In the Ciliphoroa, *Anoplophrya* forms chains of cells, whilst *Carchesium* and *Zoöthamnion* form branching colonies (Figs. 13 and 14). In *Zoöthamnion* the myonemes run throughout the length of the colony so that if one part contracts then all the rest contracts. In *Carchesium* the myonemes are restricted to each unit so that they

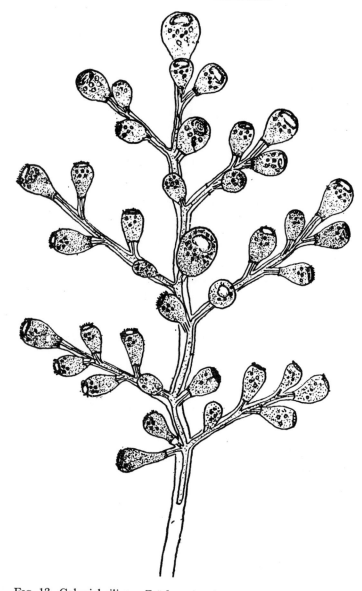

FIG. 13. Colonial ciliate. *Zoöthamnion* shows a differentiation of its parts so that there are feeding zoids and reproducing zoids. In many respects it appears similar to a colonial coelenterate. (From Hyman.)

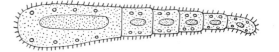

FIG. 14. Colonial ciliates. *Anoplophrya* is colonial only in that the cells formed by asexual division often remain attached in the form of a chain. (From Borradaile and Potts.)

contract individually. *Zoöthamnion* is of interest in that it shows considerable variation in the structure of its units, the colony showing division of labour (Fauré-Fremiet (1930); Summers (1938)).

Colonies are found in the cnidosporidian Sporozoa, the spores showing well-marked differentiation into cnidocysts each with its own nucleus, a spore nucleus and a spore case nucleus (Fig. 15). Whether it is justifiable to regard these reproductive units as

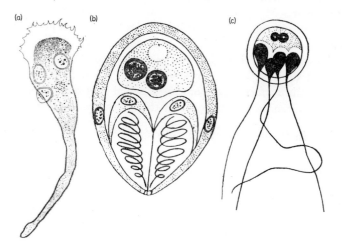

FIG. 15. Colonial sporozoa. The spores of the Cnidosporidia have a complex structure. (a) Shows the adult trophozoite and it will be seen to resemble the trophozoite of other sporozoans such as *Monocystis*. (b) Spore case together with the undischarged thread cells. (c) Spore case with discharged thread cells.

     (a) *Sinuolenea*. From Grassé after Davis.
     (b) *Myxobolus*. From Grassé.
     (c) *Chloromyxum*. From Grassé after Kudo.

colonial forms is not clear.   Baker (1948) and Ulrich (1950) suggest that the Cnidosporidia may be degenerate Metazoa but since complex spores are also found in other Sporozoa, e.g. the cysts of gregarines, and since the sporozoite of the Cnidosporidia is very similar in structure to that of *Monocystis*, it is more likely that they are real sporozoans and not degenerate Metazoa (Fig. 16).

Under certain conditions the rhizopod *Naegleria* can aggregate so that the cells being joined by a sticky material form a sheet of tissue (Willmer 1956). *Sphaerozoum* is another colonial rhizopod —a radiolarian—and in the Mycetozoa there are many forms that show colonial structure at certain stages of their life history. What is of interest here is that colonies such as *Dictyostelium* have developed a chemical system that keeps the amoebae that go to make up the colony in a unit (Bonner 1949). In general the Rhizopoda tend to form syncytia more easily than they form colonies. On the other hand quite complex multilocular skeletons are found in the Foraminifera, but there is usually only one living cell present.

From the foregoing account it is evident that the Metazoa *could have* arisen from the colonial protozoans. There are a large number of colonial Protozoa and many of them show differentiation and division of labour amongst the colony. Whether the Metazoa did *in fact* arise from the colonial protozoans is another matter and we must now consider the alternative theories.

(2) Origin from a syncytial cell

This theory suggests that the origin of the Metazoa must be sought from a protozoan that had many nuclei and which later developed membranes separating these nuclei off. The syncytium differs from the colony in that the primary unit is the whole animal and that it later becomes multicellular. In the colony the primary unit is the cell and many of these units come together to form the animal.

The differences between the syncytium and the colony are not as clear cut as could be desired. In the main they depend upon the absence of cellular boundaries. But what does one say in the case of *Volvox* or *Pheopolykrikos* where there is protoplasmic connexion between the cells (Figs. 10 and 11). Is this a multicellular animal or a syncytium?

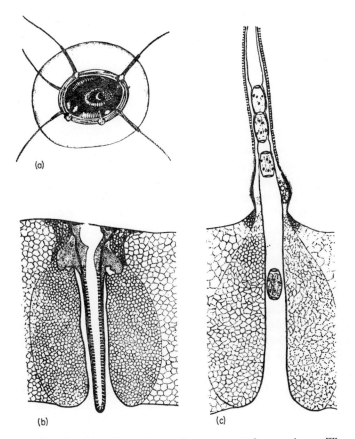

(a)

(b)   (c)

FIG. 16. Sporozoan structure. Spore cases of gregarines. The
Cnidosporidea are not the only forms to have complex spore cases;
the gregarines have them too. (a) shows the complete spore case.
(b) shows details of the tube through which the spores are discharged. (c) shows the tube everted and the spores being discharged
through it. Such a reproductive spore case with its evertible tubes
has something in common with the spore case of the Cnidosporidea.
(a) *Gregarina munieri* after Schneider. (b and c) *Gregarina ovata*
after Schnitzler. (Both from Grassé.)

Baker (1948) suggested that since in his definition a cell is " a mass of protoplasm largely or completely bounded by a membrane and containing within it a single nucleus formed by the telophase transformation of haploid or diploid set of anaphase chromosomes," that the Ciliophora and the Radiolaria are not cells. They contain more than one nucleus and therefore are syncytia. Other syncytia are found in the Flagellata (*Calonympha*, *Giardia*), Rhizopoda (*Sappinia*, *Plasmodiophora*), Sporozoa (*Myxobolus*) and Ciliophora (*Paramecium*).

The advantage of deriving the metazoan from a syncytial protozoan instead of a multicellular one is that in a syncytium such as *Calonympha* or *Opalina* the animal has an already established symmetry and an antero–posterior axis. All it has to do is super-impose cell walls on the established pattern. In the development of the multicellular form from the colonial pattern one has a series of units each with an already established axis and these axes have to be amalgamated and altered till the cells form a single unit.

This view of the syncytial origin of the Metazoa is supported by de Beer (1954), who writes, " there are the gravest objections to the view that the Metazoa were evolved by aggregation of separate protozoan individuals. This may have happened in the sponges and, indeed, is the most likely explanation for the lack of co-ordination, integration and individuality found in those animals. One of the most important features in the acquisition of individuality in organisms is axiation and integration throughout the body. The only way in which this can be imagined as having occurred in the transition from Protozoa to Metazoa is by means of internal subdivision of the protozoan cell, by cellularisation. Nor is it difficult to imagine how this might have been brought about, since there are Protozoa such as the Ciliate Infusoria, Haplozoa and some Sporozoa which possess many nuclei, and it would only be necessary to separate these by cell walls in order to obtain the required organisation for the primitive Metazoa."

One difficulty comes when we consider whether the syncytium has any inner cell walls or not. Thus if it can have cell walls which do not divide the parts completely, then animals such as *Volvox* which has connexions between the adjacent cells are syncytia (Fig. 10). It is true that it is possible to consider that *Volvox* has arisen by the accumulation of *Chlamydomonas* like individuals whilst it is not possible to find any sub-unit for *Opalina*. Nevertheless the differences between syncytia and colonies are not as clear cut as has sometimes been supposed.

Another difficulty arises when we consider the stage during the life cycle during which the protozoan is syncytial or colonial. Thus many of the Protozoa form spores during reproduction and these may be localised in spore cases. Are these to be regarded as a multicellular stage? If so, then the palmella stage of *Chlamydomonas* could be a colonial form (Fig. 8). The Cnidosporidia have many nuclei only during the reproductive (spore-forming) stage; their trophozoite is unicellular and only has one nucleus. Equally well the ciliate *Anoplophrya*, which does not separate its asexually produced cells from the parent immediately they are produced, can be considered as a colonial form (Fig. 14). In effect the situation is quite difficult to resolve and depends to a large extent on the relative duration of the multicellular stage and the part that it plays in the life of the animal. Hadži (1953) has suggested that one of the major differences between the Protozoa and the Metazoa is that the Protozoa have their major phase in the reproductive stage whilst the Metazoa have their major phase in the vegetative stage.

(3) Origin from the Metaphyta

The third view concerning the origin of the Metazoa is that they arose from the plants, the Metaphyta. It has already been mentioned that Franz (1926) thought that the Protozoa were in fact derived from the Metaphyta and the Metazoa. Baker (1948) suggested that the Metazoa arose from plant-like protozoans. " The unicellular plant absorbs nutriment from all sides equally, and when in the course of ontogeny or phylogeny it becomes a metaphyte there is no fundamental change in this respect; a cell divides without separation and the two products continue to absorb nutriment over most of their surface. The passage from unicellular form to the metaphyte is therefore easy. In the case of animals, however, there is an important change when a unicellular form becomes a metazoön; a new method of feeding must be adopted. . . . The difficulties would be greatest when the protozoön had a localised mouth. If the products of such an animal were to adhere together and each were to acquire its own mouth, no advance could be made to the evolution of a metazoan alimentary canal. This suggests that the Metazoa may have arisen from primitive Protozoa unprovided with localised organs of assimilation."

Hardy (1953), following on from Baker's argument, suggests that " the Metazoa have not been derived from the Protozoa at all but from relatively simple metaphytes, which after they had evolved from the protophytes began, perhaps as a result of a shortage of phosphates or nitrates, to capture and feed on small organisms as do the higher insectivorous plants."

This is an interesting suggestion but one that can be criticised on several grounds. There is no evidence that metaphytes such as the Algae can withstand food shortages by catching animacules. The thick cellulose cell wall around the metaphytes, though a protection, would also tend to prevent them from developing pseudopodia rapidly enough to catch protozoans. The insectivorous plants, it will be remembered, all develop special insect-catching mechanisms, and even so they still retain their photosynthetic ability.

Baker's arguments in favour of the origin of Metazoa from plant-like protozoans can also be contested. There is no reason why a metazoan type of alimentary canal should develop in the first stages of the evolution of the Metazoa. The " alimentary canal " of the sponges is not really comparable in function to that of the higher metazoans, and even in the coelenterates there is a considerable amount of amoeboid activity in the gut cavity. What would appear to be more important than the development of an alimentary canal is that the cells of the body should have some continuity and interconnexion with each other so that food material can be passed easily from one cell to the other. Such a process probably does occur in the colonial ciliates such as *Zoöthamnion* which have well-developed gullets. Summers (1938) suggested that food material was probably passed along the stalk of *Zoöthamnion*. Here then we have the case of a protozoan with a well-defined mouth forming a colony. Furthermore this colony shows differentiation and division of labour, some of the polyps being more intensive feeders than others.

There is no reason why a protozoan with a definite polarity should not lose this polarity and develop into a colonial form.

Willmer (1956) in discussing the change of form of *Naegleria* points out that the rhizopod can be in any one of three phases:

(1) in the flagellate stage with a definite polarity;
(2) in the amoeboid phase with pseudopodia coming from all over the body;
(3) aggregated in the form of a sheet of tissues.

It would seem that this protozoan has no difficulty in losing its polarity and therefore that the difficulties raised by Baker concerning the changes from Protozoa to Metazoa are not as great as he suggests.

What conclusion then can be drawn concerning the possible relationship between the Protozoa and the Metazoa? The only thing that is certain is that at present we do not know this relationship. Almost every possible (as well as many impossible) relationship has been suggested, but the information available to us is insufficient to allow us to come to any scientific conclusion regarding the relationship. We can, if we like, *believe* that one or other of the various theories is the more correct but we have no real evidence.

CHAPTER 6

THE MOST PRIMITIVE METAZOA

WE HAVE seen so far that the Metazoa can be derived either from
syncytial protozoans, from multicellular protozoan colonies or
from the Protophyta. To some extent the theory that one chooses
as the most probable will depend upon which group is considered
to be the most primitive of the Metazoa. Thus if one considers the
Sponges as the most primitive of the Metazoa then one could
suggest a link between the Protozoa and the Metazoa via the
Choanoflagellata. If on the other hand one thinks that the
Acoelous Platyhelminthes are the most primitive metazoans, then
one could consider that the link with the Protozoa was via the
complex ciliates. It is therefore important to decide which are the
most primitive of the Metazoa, but before this can be done one
has to consider four questions.

(1) Which of the metazoan groups can be considered the
earliest to have evolved?
(2) Which are morphologically the most simple of the Metazoa?
(This will not necessarily be the first group to have evolved.)
(3) What is the relationship between the major groups of the
lower Metazoa?
(4) Can the Metazoa be considered as a polyphyletic group
with more than one origin from the simpler living forms?

The metazoans that will be considered here are the five groups:
(1) Porifera.
(2) Mesozoa.
(3) Coelenterata.
(4) Ctenophora.
(5) Platyhelminthes.

Before discussing these questions it will be as well to indicate

50

the various uses of the expressions " primitive; simple; advanced; radial symmetry and bilateral symmetry," since these terms will frequently be used in the following discussion.

Primitive and simple

There are two terms that must be distinguished and used carefully. The first term is " simple." If an animal has a morphological structure made up from a few basic units, then such an animal can be regarded as having a simple structure. Other animals may have many different units arranged in a variety of patterns; they can then be regarded as having a complex structure. These two groups, the simple and the complex, can also be described as having a low level of complexity or a high level of complexity.

The second term is " primitive." This means that of two structures or conditions, one arose some time before the other. The concept of "time of origin" is the critical point in determining whether a structure is primitive or not. Because an animal has a simple morphological pattern it does not mean that it had an early evolutionary origin and therefore is in a primitive condition.

It is perhaps unfortunate that during most courses of Zoology the students are taken from the Protozoa to the Primates and shown the way in which the complexity of structure increases. Quite often the student becomes puzzled when he deals with the Mollusca. Should they come before the annelids, between the annelids and the arthropods, or after the arthropods? It is clear that this problem confuses two issues: firstly the complexity of molluscs in relation to that of the annelids and the arthropods, and secondly the time of origin of the molluscs, i.e. did they arise before or after the annelids?

The student is usually taught that certain conditions can lead to a simplification of morphological form and that clues other than purely morphological ones must be used to elucidate an animal's phylogenetic position. In particular this holds when we come to deal with parasitic animals. Thus the larval form of *Sacculina* quite clearly shows the crustacean ancestry of the parasite even though the morphology of the adult is not at all typical of the Crustacea. Though the parasitic habit is usually associated with certain morphological changes, the other specialised ecological conditions are not often given equal credit for determining and shaping an animal. Thus various morphological conditions will be associated with a pelagic life, with burrowing, with living in sand, with being a very large animal or being a very small animal. All of these tend to alter the morphology of the animal and make it a successful living animal, not just a representative of a hypothetical idea; that of, say, a crustacean. An example of such an environmental effect can be seen when we come to consider which is the more primitive, radial symmetry or bilateral symmetry? (Fig. 17.)

Radial symmetry

This type of symmetry is often found in sessile or pelagic animals. They usually have an oral and an aboral surface but otherwise any diameter cut at right angles to the oral–aboral axis should divide the animals into two similar halves. In fact most of the animals that are radially symmetrical do not fit in with this definition since they usually have some irregularity in their organisation, i.e. mesenteries, tentacles, madreporite, which allow only certain sections at right angles to the oral–aboral axis to divide the animal into equal halves.

Bilateral symmetry

The animals that show bilateral symmetry are organised into an antero–posterior axis and a dorso–ventral axis. In addition there is one plane and one plane only that will separate the animals into equal right and left halves. A radially symmetrical animal will have many such planes. Most of the Metazoa that are not pelagic or sessile show a bilateral symmetry. (Fig. 17.)

One is often taught that the coelenterates and the echinoderms show a basic radial symmetry and that the other Metazoa are bilaterally symmetrical. Since the echinoderms and coelenterates are sometimes placed at the foot of the metazoan evolutionary tree it is not difficult to associate radial symmetry with a primitive habit and to assume that the bilaterally symmetrical condition is the more advanced. On the other hand, it is equally true that radial symmetry is found in sessile or floating animals whilst bilateral symmetry is found in crawling or swimming animals. We are therefore left with the question, "To what extent does the symmetry of an animal indicate its primitiveness and to what extent does it reflect the habits of that animal? "

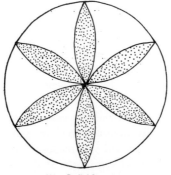

(A). Radial Symmetry

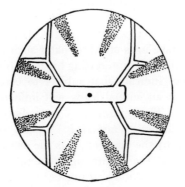

(B). Biradial Symmetry

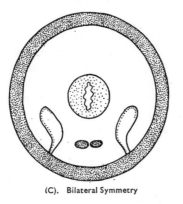

(C). Bilateral Symmetry

FIG. 17. Types of symmetry.

There are certain exceptions to the generalisation that sessile animals are radially symmetrical. Thus, as previously mentioned, even amongst the coelenterates there are many planes that will not divide the animal into two equal halves, this being due to the development of tentacles, gonads, batteries of nematocysts, mesenteries and siphonoglyphs. In other Metazoa it is rare to find radial symmetry. Thus in the Rotifera, neither *Trochosphaera* nor *Melicerta* are perfectly radially symmetrical. In the Annelida *Sabella* is not radially symmetrical; its parapodia still show a bilateral symmetry. In the barnacles, though there is some tendency towards a radial symmetry as illustrated by the skeletal plates, the internal symmetry of other organs such as the legs, digestive system and nervous system is a bilateral one. Other sessile animals such as the Crinoids, Ascidians and Pterobranchiates do not show perfect radial symmetry. On the other hand the ctenophores that take up a crawling habit such as *Coeloplana* and *Ctenoplana* do show a very interesting bilateral (biradial) symmetry.

These examples indicate that subject to certain basic limitations, the life that an animal leads will influence its shape and basic symmetry. The question, " Is radial symmetry more primitive than bilateral symmetry? " should perhaps be more correctly replaced by the question, " Is the sessile or pelagic habit more primitive than the swimming and crawling habit? " The answer to the latter question is at present unknown.

(1) The Sponges (Porifera)

The sponges are peculiar multicellular animals with an organisation quite different from that of the other Metazoa. They have a skeletal system and three layers of cells, pinacocytes, amoebocytes and choanocytes, but they have no organ systems such as an excretory or a nervous system. They have a very simple digestive system in which there is no real mouth or gut. The sea water around the animal passes through a series of apertures into the centre of the sponge and in doing so is filtered, the food being taken up by the choanocytes and the amoebocytes.

The organisation of the sponge is very simple in that a sponge can be passed through the meshes of a net and so separated into its individual cells. These cells can later aggregate and form an organised sponge with the cells in their correct relative position;

the mechanism of this interesting rearrangement is not yet understood. This type of cellular organisation is found in the higher Metazoa where it has superimposed on it the co-ordinating influence of a nervous and hormonic integration, both of which are apparently absent in the sponges. The high degree of skeletal material relative to the small amount of living protoplasm makes the sponges very poor food and so a relatively successful group of animals.

Are the sponges a primitive or an advanced group of animals? To answer this question we should have to know the time of origin of the sponges and there is no certain information on this point. Instead we can examine the apparently simple characters and the apparently complex characters and attempt to derive some satisfaction from this. The reader should always be on his guard when consulting such lists; it is not possible to come to a conclusion merely by seeing which of the two lists is the longer!

Simple characteristics

    (1) The layers of the body are loosely organised.
    (2) The layers of the body do not correspond to the ectoderm, mesoderm and endoderm of the higher forms.
    (3) There is no definite body form.
    (4) They have choanoflagellate cells like those present in the choanoflagellate protozoans.
    (5) There is no nervous system.
    (6) There is no excretory system.
    (7) There is no mouth.
    (8) The gut (gastral cavity) shows little differentiation.
    (9) They have a high regenerative capacity.
  (10) They have a well-developed system of asexual reproduction.
  (11) The larvae have well-developed flagella.

Complex characteristics

    (1) They have three layers of cells (some coelenterates have only two layers of cells).
    (2) They have a well-developed middle layer, the " mesenchyme," with an elaborate skeletal system.
    (3) They have some differentiation within the layers, e.g. the pore cells.

(4) They have a gut. (The Mesozoa have no gut.)

(5) The gemmules with which some sponges carry out asexual reproduction are quite complex in structure.

(6) They have eggs and sperm.

(7) The embryo gastrulates in a complex manner, by ingression, epiboly and delamination.

(8) They have a well-developed amphiblastula larva.

(9) The blastopore is aboral, in most other metazoans it is oral.

(10) An inversion of the layers occurs during embryology.

(11) The sponges are the only animals to have the main body aperture the exhalant one.

Various excuses can be made for each and every one of the simple or specialised characteristics. Thus, to deal with a few of them:

(1) Though it is correct that the layers are loosely organised this may be due to the fact that the sponges do not depend on the hydraulic pressure of the gastro-vascular cavity to maintain their shape. They have a well-developed skeletal system that takes care of this. The animal remains intact even though there are many series of canals running through the body, and the cells are not firmly cemented together.

(2) The layers do not correspond to the ectoderm, endoderm and mesoderm of the higher animals and there is no evidence that at one time they did correspond. There is not one piece of evidence to show that the normal triploblastic condition evolved from that shown by the sponges.

(3) Many sponges such as *Euplectella* and *Poterion* do have a definite shape.

(4) Choanoflagellate-like cells are also found in the endoderm of some coelenterates, annelids and molluscs. Electron microscope studies show that the collar is a series of protoplasmic filaments that project out of the cell in much the same way that the digestive filaments project from an endodermal cell (Rasmont *et al.* 1958).

(5) There are other animals that have no nervous system, e.g. the Mesozoa.

(6) The Mesozoa, Coelenterata, Ctenophora and Acoela have no specialised excretory system.

(7) The gut in the sponges though it shows little differentiation is at least a gut. There is no gut in the Mesozoa, Acoela or Pogonophora.

(8) A high regenerative capacity does not indicate that an animal is simple. Thus in the coelenterates, the polyps have good regenerative capacity, the medusae do not. In the annelids the Oligochaeta have good regenerative capacity, the Hirudinea do not. In the amphibians the Urodeles regenerate well, the Anura do not.

(9) There is no evidence that asexual reproduction is necessarily more primitive than sexual reproduction.

It can be seen that the situation is not as straightforward as the lists might at first sight make it appear. It should also be remembered that though one might believe that the sponges are more simple than the coelenterates from the point of view of their morphology and life cycles, this is no reason for thinking that they necessarily developed some time before the coelenterates. The sponges may have been a happy afterthought of the Protozoa after they had given rise to the coelenterates!

Which group of animals did the sponges come from? There are three different answers to this question. The first derives the sponges from Protozoa such as the choanoflagellates or Volvocinae. The second derives the sponges from a Gastrea type of animal which gave rise to both the coelenterates and the sponges. The gastrula probably came from a colonial protozoan such as *Volvox*. The third answer derives the sponges from a coelenterate source. Each of these views has something in its favour and something against it.

ORIGIN FROM THE PROTOZOA

(1) From the Choanoflagellata

The inner layer of cells in the sponges is composed mainly of Choanocytes. These in many ways resemble the cells of the choanoflagellate protozoa and on this basis it has been suggested that the sponges might be derived from these protozoans.

In 1880 Saville Kent described a colonial choanoflagellate called *Proterospongia* (Fig. 18). This consisted of a flat plate of about forty cells. The other cells had the normal choanoflagellate structure whilst the inner cells were amoeboid. Periodically a choanoflagellate cell would withdraw its flagellum and become amoeboid, whilst an amoeboid cell would take up the position and

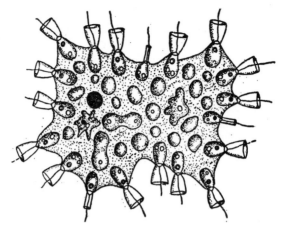

(A). Proterospongia

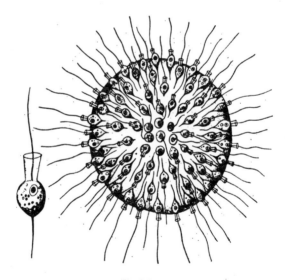

(B). Sphaeroeca

Fig. 18. Colonial protozoa that in some ways resemble sponges.
(A) *Proterospongia*. (After Saville Kent.) This is now believed
    to be a fragment of a fresh-water sponge.
(B) *Sphaeroeca;* single choanoflagellate protozoan and the colonial
    form. (From Grassé after Lauterborn.)

structure of a choanoflagellate cell. This protozoan had in many ways the structure that one might expect to find in an ancestral sponge and it has usually been figured as such in textbooks of zoology.

Proterospongia is not a common protozoan. Recently Tuzet (1945) has investigated the antecedents and morphology of this protozoan and she decided that in fact Saville Kent was the only person ever to have definitely seen *Proterospongia* and that what he saw was not a protozoan but a small fragment of an actual sponge. Tuzet concludes, " Pour nous. . . . La *Proterospongia* de Saville Kent n'est pas autre qu'un corps de restitution d'Eponge d'eau douce." *Proterospongia* is nothing more than a restitution body of a fresh-water sponge. If this is true it is not surprising that *Proterospongia* has many sponge qualities. On the other hand Gröndtved (1956) has described a new species of *Proterospongia*, *P. dybsoensis*, in which there are three to ten cells arranged in a linear row per colony. These colonies were often found in very large numbers, up to 2,300 colonies per litre of water, and Gröndtved thought that they might be fragments from a larger colony except for the fact that they were all so much alike. There does seem to be quite a considerable difference between the row of three to ten choanoflagellate cells embedded in a gelatinous common envelope and *Proterospongia* as described by Saville Kent, where the choanoflagellate cells migrated into the interior of the colony and took up amoeboid structure. For this reason it is not clear whether the new species rightly can be placed in the genus *Proterospongia*.

(2) From the Volvocinae

There are certain resemblances between the embryonic development of *Volvox* and that of certain sponges. Duboscq and Tuzet (1937) showed that in *Grantia* the embryo developed inside a membrane. The blastula is made up of two types of cells, flagellate ones and non-flagellate ones. In the blastula all the flagellate cells point inwards at first but during the course of development the blastula turns inside out so that the flagella now point outwards. This phenomenon is called inversion and is shown diagrammatically in Fig. 19. The larva is then liberated as an amphiblastula larva with the flagella at one end. The other

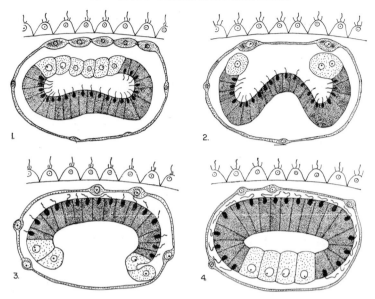

FIG. 19. Diagram to show inversion during the embryology of the sponge *Grantia*. The embryo develops at first with its flagella pointing inwards (1). The embryo slowly inverts (2–3) so that its flagella now point outwards. (After Duboscq and Tuzet.)

end of the larva has the non-flagellated cells and after swimming for some time the amphiblastula larva settles on the substratum, the non-flagellated cells grow over the flagellated cells so that the animal takes up the structure of an adult sponge.

A similar inversion takes place during the development of the daughter colonies of *Volvox* (Pocock 1933). During the asexual development of a daughter colony the daughter cells divide and form a sheet of cells. These cells are orientated in the same manner as the parent cells, the flagella pointing outwards, but as division proceeds the daughter cells form a ball with the flagella pointing towards the centre of the ball. The colony is not a complete ball since there is a hole at the top. The colony now proceeds to turn itself inside out through this hole in much the same way as one might push a tennis ball inside out through a hole in the wall (Fig. 20). This results in a colony with the flagella all pointing outwards, the inversion taking some two hours.

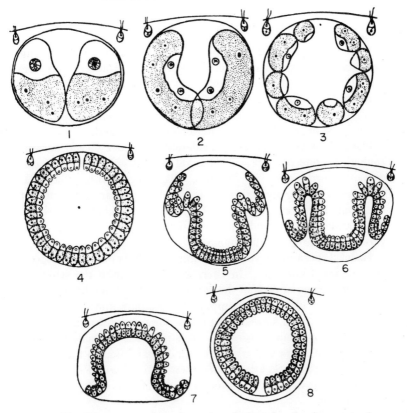

FIG. 20. Diagram to show inversion during the development of daughter cells in *Volvox*. The flagella of the daughter colony all point towards the inside of the colony. Inversion takes place and the colony turns inside out so the flagella now point outwards. (After Zimmerman.)

In both *Volvox* and *Grantia* the cells develop in the same initial relationship to the embryo as they do to the parent layers. Hyman (1940) was impressed by the similarity between the inversion in *Volvox* and the sponges and suggested that this might indicate a common ancestry. On the other hand Tuzet (1945) considered the situation as one of convergence which does not indicate any underlying phylogenetic relationship.

ORIGIN FROM THE GASTRULA

This view holds that the sponges evolved from some form of gastrula and was propounded mainly by Ernst Haeckel. Haeckel from his studies of the embryology of the sponges (1872) decided that the larval form of the sponge was a gastrula larva. He also thought that certain other animals such as *Haliphysema* were primitive sponges, though he later changed his mind.

Haeckel's views have had considerable influence on our current zoological concepts. It was Haeckel who devised many of our current words such as Phylum, Blastula, Morula, Gastrula, Ontogeny, Phylogeny and many more. It is perhaps relevant that we should spend a little time reporting how Haeckel developed his ideas and concepts and the way in which these fitted in with views on the origin of the sponges.

Haeckel spent many years developing his views on the phylogeny of the animal kingdom. In effect he studied both the structure and embryology of each of the various groups. Then by marrying the facts and ideas from comparative anatomy and embryology, Haeckel developed a sweeping plan of the relationship and evolution of various animal groups. He supported his schemes with detailed arguments and when his opponents failed to understand his arguments Haeckel devastated them with a barrage of crushing sarcasm against their misinterpretation of his own specialised terminology.

Simply stated, Haeckel's view was that the most primitive animal was a small non-nucleated mass called the Monerula (Fig. 21). This was followed by a later group of animals that had a nucleus and this state was called the Cytula. The Protozoa are at the cytula stage. The next group of animals formed a solid mass of nucleated cells called the Morula. The Morula led to a more complex form that had a hollow centre and a single layer of cells, the Blastula. Certain cells such as *Volvox* are almost at the Blastula stage. The next stage in evolution was the Gastrula stage, in which the animal had a double wall, a ciliated exterior and a hollow gut.

The Gastrula assumed tremendous importance in Haeckel's phylogenetic speculations. He thought that the Gastrula was the ancestor of all the Metazoa, that it occurred in all the Metazoa at

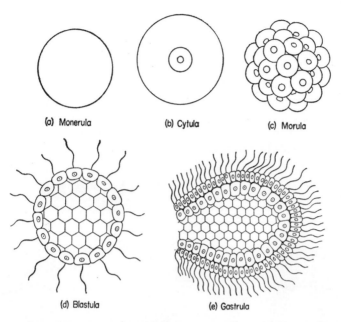

(a) Monerula   (b) Cytula   (c) Morula

(d) Blastula   (e) Gastrula

FIG. 21. Haeckel's concept of levels of organisation. He suggested that animals evolved through the successive adult stages shown in the above figure.

some stage of their embryonic development and that a group of animals existed which were adults but which were still at the gastrula stage. Such animals were not known, but later Haeckel thought he discovered such an adult group of animals and he called them the Physemaria, an example of which was *Haliphysema*.

Haliphysema, according to Haeckel, had the structure of a little vase. The walls of this vase were made up of two layers of cells; the inner layer was flagellated, the cells having a collar-like structure. The outer layer was made up from a syncytium of cells. These outer cells took up stones and spines from the environment and covered the animal with a protective layer. They were adults, since Haeckel described a series of gonadial cells which were formed from the endoderm and then became liberated into the gut (Fig. 22). Many genera related to *Haliphysema* were discovered and they all showed certain resemblances to sponges, i.e. a central cavity lined with flagellated cells, and an outer layer of cells covered by a skeleton which in this case differed from that of a sponge in that it was not secreted but picked up from the substratum. These animals assumed such tremendous importance to Haeckel that at one time he derived all the Metazoa from the Physemaria, but at a later date he decided that the Physemaria were on a side line from the main Gastrula.

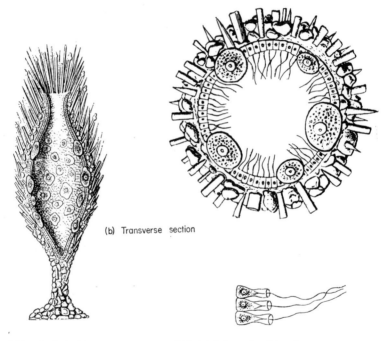

(b) Transverse section

(a) Longitudinal view   (c) Isolated choanoflagellate cell

Fɪɢ. 22. Haeckel's view of the structure of *Haliphysema*. He thought that the structure was that of a simple sponge. (a) shows the whole animal with part of the body cut away to demonstrate the inner structure. (b) is a transverse section of *Haliphysema* and shows the gonads migrating into the interior of the animal. (c) shows in more detail the structure of the inner flagellate cells. It is doubtful if such a detailed pattern exists in *Haliphysema* or if such a simple condition exists in any sponge.

A difference of opinion arose between Haeckel and other workers over the structure of the Physemaria. Saville Kent in 1878 from studies of both living and fixed material decided that *Haliphysema* was no sponge but instead a foraminiferan rather like *Euglypha*. He was unable to see any of the internal details described by Haeckel. A controversy soon arose between Haeckel and Saville Kent and it was left to Ray Lankester, as a friend of Haeckel, to enter the controversy in the role of adjudicator.

Lankester asked Saville Kent for specimens of *Haliphysema* and then examined them both alive and in the fixed and stained condition. After considerable examination Lankester (1879) decided that Saville Kent was perfectly correct in his assertions and that the specimens were clearly those of a foraminiferan. But the matter did not end there. Lankester ingenuously decided that the answer to such a controversy was extremely simple. He suggested there must be two different genera of animals which from the outside looked *exactly* alike but one of these had been studied by Professor Haeckel whilst the other had been studied by Mr. Saville Kent. Lankester had no doubt that the isomorph studied by Haeckel would have the structures that Haeckel had described and he hoped that Professor Haeckel would supply him with some specimens. It does not appear that such specimens were ever sent to Lankester.

Perhaps something should be said in Haeckel's defence. In a recent paper on *Haliphysema tumanowiczii*, Hedley (1958) describes the way in which the protozoan often picks up sponge spicules and covers itself with these. Also certain individuals were multinucleate, a condition which in certain circumstances might be confused with some of the conditions described by Haeckel.

Haeckel continued to believe in the importance of the Physemaria though he thought that the Gastrula was more important (1899). He derived all the Metazoa from the Gastrula and stated, " I regard the Gastrula as the most significant and important embryonic form in the whole animal kingdom. It occurs amongst the sponges, Acalephe, the Annelida, Echinodermata, Arthropoda, Mollusca, and the Vertebrata as represented by *Amphioxus*. In all these representatives of the most various animal stocks, from the sponges to the vertebrates, I deduce, in accordance with the Fundamental Biogenetic Law, a common descent of the whole animal world from a single unknown stock form, Gastrula or Archigastrula, which was essentially like the gastrula." (Monograph on Calcispongiae.)

These ideas were not accepted even in Haeckel's own time. Thus both Claus and Schmitt disagreed with him, and as Radl (1930) states, " The popular idea of the method of the scientist is that he assembles a series of definite facts upon which he founds his case. We see that this is not always the case. It is not true that the facts which told against the Gastrula were unknown at the time when the theory was propounded, or that the theory was gradually discredited as the facts which contradicted it were gradually accumulated until it finally had to be abandoned. Everything that has ever been cited against the theory was known when the theory was put forward; nevertheless it was widely accepted. Today some still accept it, others do not." Though it is perhaps an overstatement that " everything that has ever been cited against the theory was known when the theory was put forward," Radl's point is made quite clear. The theory was often accepted because it was attractive and not because it was supported by detailed verified factual information.

One may conclude, therefore, that there is at present no evidence that the sponges arose from an adult gastrula. They have neither a hollow gastrula larva nor are there any simple sponges that are still in the gastrula condition during the adult stage. The situation as Haeckel saw it was based on over-simplification and misinterpretation of the evidence.

A modification of the Gastrula theory has recently been proposed by Jagersten (1955). He suggested that the primitive blastula gave up living and swimming in the sea and started to crawl on the sea bottom (Fig. 23). It modified its structure to become a " Bilateroblastea " with a flattened ventral surface, an arched back, a few sensory cells at the front of the body and the sexual cells inside the body. The centre of the body was hollow and not filled with mesodermal cells. At first the animal fed phagocytically all over the body surface, but as food particles accumulated on the ventral surface the phagocytic ability became restricted to the ventral region. This then became raised from the ground till the animal took up the shape of the " Bilaterogastrea " as shown in Fig. 23.

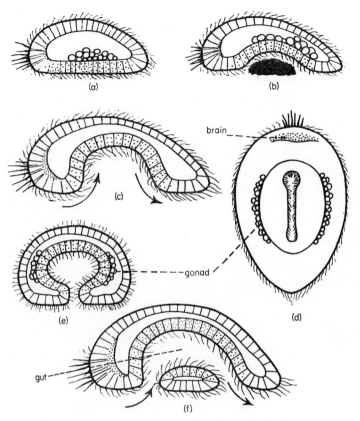

FIG. 23. Bilaterogastrea theory. Jagersten has suggested that the primitive larval form was that of a bilaterogastrea. The planula larva (a) settled on the ground, raised its ventral surface to accommodate food material (b, c) and so developed a gut (d, e, f). It then takes up the form of a Bilaterogastrea. (From Jagersten.)

Jagersten's scheme is derived from his view that the primitive form had a hollow centre and was not filled with mesenchyme. The reason he thinks that the primitive form had a hollow interior is that otherwise the various groups of animals such as the coelenterates, platyhelminthes, etc., would have had to develop a hollow gut independently and on several different occasions. Secondly, though he agrees with the statements that the endoderm

6—IOE

in the Hydrozoa is formed by the wandering in of cells, in the Scyphozoa, Anthozoa and higher animals the endoderm is often formed by invagination. Since on other grounds Jagersten thinks that the Anthozoa are more primitive than the Hydrozoa, he suggests that invagination is the more primitive system in the formation of endoderm and that the primitive larva had a hollow centre, i.e. a gut, and was not solid as has been suggested by Hyman (1940).

The following system is therefore suggested by Jagersten to explain the development of the Porifera from the bilaterogastrea. The bilaterogastrea settled on the sea floor and placed the middle of its elongated mouth on the substratum. The water and food material flowed in through the mouth and out via the anus. The mouth later became folded and developed a series of pores as shown in Fig. 24. The anus remained a single structure and migrated to a dorsal position to become the exhalant opening. The animal then had the form of a sponge though it would still have to develop the peculiar histological structure of the Porifera. Jagersten's view is of interest in showing what *could* have happened. Whether the Porifera did actually arise in this manner is open to doubt.

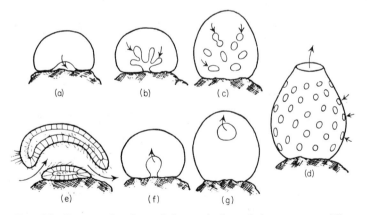

FIG. 24. Jagersten's view of the evolution of the sponges. The bilaterogastrea settled on the bottom (a), raised its mouth from the substratum (b) and then divided the mouth into many oscula (c). The anus then migrated dorsally (f, g). (e) is a transverse longitudinal section of a.

ORIGIN FROM THE COELENTERATA

This view, that the coelenterates and the Porifera have close ancestral affinities is quite an old one. It is based on the fact that the coelenterates and the sponges often have a solid planula-type larva during their embryology. Lankester (1890) strongly supported this view when he emphasised the differences between the planula larva and the Gastrula. He thought that there was no indication that the sponges ever had a gastrula stage but that instead the resemblance was in the solid blastula.

In *Leucosolenia* the fertilised egg divides to form a sixteen-celled hollow blastula. The majority of the cells are flagellated but a few at one end are non-flagellated. These non-flagellated cells together with a few of the flagellated cells migrate into the interior of the blastula and fill the central cavity. The result is a solid blastula. This settles on the ground, flattens and the inner cells then migrate out on top of the flagellated cells. They then become the pinacocytes and amoebocytes, whilst the flattened flagellated cells turn into the choanocytes.

Though there are certain similarities between the development of the sponges and the coelenterates, it is difficult to know how much reliance can be placed on them. Thus Balfour (1880) was quite clear that there was no relationship between the cell layers of the sponges and those of the coelenterates since the inner layers of the sponge embryos come to lie on the outside of the adult. For the same reason Delage (1898) suggested that the sponges have their endoderm on the outside and the ectoderm on the inside and that the sponges should be called the " Enantiozoa " for this reason. It is of interest that Saville Kent (1880) described the way in which the outer cells of the sponge, the pinacocytes, could take in food particles.

It is possible to enumerate the similarities and differences between the sponges and the coelenterates as follows.

Similarities between sponges and coelenterates

(1) They are both aquatic and free-living animal groups.
(2) They have spicules in their skeleton, which are either calcareous or horny.
(3) They have flagella.

(4) Amoebocytes are present in both groups.
(5) The main body cavity is neither a haemocoel nor a coelom.
(6) There are no excretory organs.
(7) They occasionally have mesenchyme but never mesoderm.
(8) They form buds or gemmules for asexual reproduction.
(9) They have sexual reproduction with sperm and eggs.
(10) They both form colonies.
(11) They both have a high regenerative capacity.
(12) They have a solid blastula larva (stereoblastula) before the amphiblastula or planula larva develops.
(13) The anterior end of the larva becomes attached to the ground.
(14) The sex cells are formed by interstitial cells or amoebocytes.

Differences between sponges and coelenterates

(1) The sponges have many entrances to the body cavity.
(2) The main body opening of the sponges is the exhalant one.
(3) The sponges have choanocytes (but see p. 56).
(4) The outer layer of the sponge is never ciliated in the adult.
(5) There are no nematocysts in sponges.
(6) The sponges have no muscles or musculo-epithelial cells (except porocytes).
(7) The sponges have no nervous system or sense organs.
(8) The sponges do not show polymorphism.
(9) The sponge spermatozoa are sometimes carried to the egg by amoebocytes.
(10) The sponges never form a compact skeleton such as is seen in certain coelenterates such as the corals.
(11) The adult sponges are never pelagic.
(12) There is no clear homology between the layers of the sponges and the layers of the coelenterates.

Though the above lists do not actually prove anything they do indicate that the differences between the coelenterates and the sponges are quite considerable and basic. It is thus doubtful if there is any close relationship between these two groups. It is also impossible to state whether the sponges arose earlier than the coelenterates. Their organisation is less complex in some ways, but this again does not necessarily mean that the sponges are

therefore more primitive than the coelenterates. Our conclusion,
therefore, is that the situation is not at all clear.

(2) THE MESOZOA

The Mesozoa are a group of parasitic animals of very simple
structure. They are multicellular and usually take the form of a
solid mass of cells with one or more internal cells. These internal
cells are not digestive in function but instead play a part in the
reproduction of the animal.

The mesozoans in some ways correspond in structure to a solid
blastula and it has been suggested by some writers such as van
Beneden (1876) and Hyman (1940) that the Mesozoa are a
primitive group, or even the most primitive group, of the Metazoa.
On the other hand the Mesozoa are all internal parasites. Thus

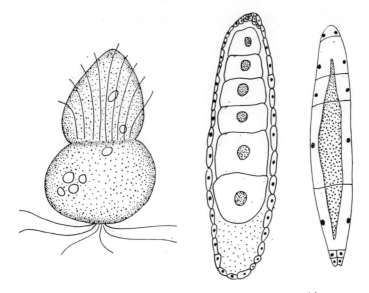

(a) Ciliated larva   (b) Female   (c) Male

FIG. 25. Mesozoan structure. *Rhopalura.*

(a) Ciliated larva. (From Hyman after Atkins.)
(b) Adult female. (From Hyman after Caullery.)
(c) Adult male. (From Hyman after Caullery.)

one group, the Dicyemida, represented by *Dicyema*, is found in the kidney of the *Octopus*, whilst the other group, the Orthonectidae, represented by *Rhopalura*, is found inside various marine invertebrates. Both these groups have a ciliated larva which may bore into a new host and this larva in some ways resembles the miracidium larva of the digenetic trematodes (Figs. 25 and 27). The Mesozoa also have a complex life cycle, and this together with the larval structure has led writers such as Stunkard (1954) and Caullery (1951) to think that the Mesozoa are probably digenetic trematodes. The recent *Traité de Zoologie* edited by Grassé also places the Mesozoa amongst the platyhelminthes. We thus have two views concerning the Mesozoa; one that they are primitive animals, the other that they are degenerate parasites.

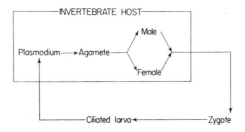

FIG. 26. Diagram of the life cycle of *Rhopalura*. (After Caullery.)

It should be stated at the very beginning of this discussion that the Mesozoa have suffered as a group in that various non-related animals such as *Haplozoön* have been thrust into the Mesozoa though in fact their affinities are elsewhere (see p. 41). There is also some doubt about the closeness of the relationship between the Orthonectids and the Dicyemids. The resemblance lies in their simple morphology and the fact that both have a ciliated larva, but since they are both internal parasites, the simple morphology is suspect straight away. One knows that other internal parasites such as the males of *Bonelia*, or the parasitic cirripedes, become very simplified. On the other hand the adult trematodes and cestodes are not morphologically simple—there being a tremendous development of the reproductive systems.

Dodson (1956) has suggested that though a parasitic life can lead to morphological simplifications in the parasite, it does not

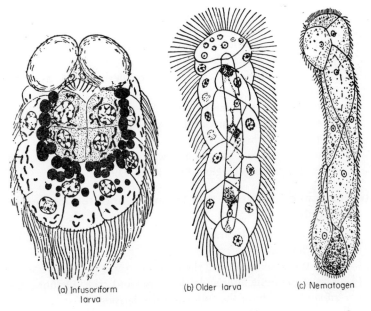

(a) Infusoriform   (b) Older larva   (c) Nematogen
larva

FIG. 27. Mesozoan structure. *Dicyema.*

   (a) From Hyman after Nouvel.
   (b) From Hyman after Lameere.
   (c) From Hyman.

necessarily do so. Thus animals such as the leech are parasitic but complex in structure, and, as we have already mentioned, many of the platyhelminthes and nematodes are quite complex. When one also considers the fact that the mesozoans have a complex life cycle, some stages of which are not yet known, it would appear premature to place the mesozoans in a key position between the Protozoa and the Metazoa.

It is possible to draw up lists of the simple characters and the platyhelminth-like characters of the Mesozoa. These are as follows.

Simple characters of the Mesozoa (non-platyhelminth characters)

  (1) They are multicellular animals with no differentiation into endoderm, ectoderm or mesoderm.

(2) They have a solid blastula.

(3) There is a simple adult form; there are no proglottides, no suckers, no thick cuticle, no nervous system, no flame cells, no complex gonadial system.

(4) They have cilia and a few reproductive cells as their specialisations.

(5) The cilia of the trematode miracidium larva are soon lost; those of the Mesozoa last throughout the life of the animal (not in the Orthonectids).

(6) There is no cell in the miracidium comparable to the internal nematogen cells of the adult Dicyemids.

Resemblances between the Mesozoa and the digenetic trematodes

(1) They are internal parasites.

(2) They have a complex life cycle.

(3) Both the trematodes and the Mesozoa show polyembryony.

(4) The trematode miracidium larva and the Orthonectid ciliated larva have the following similarities.

(a) The larva results from similar unequal cleavage of the fertilised ovum.

(b) The larva is bilaterally symmetrical.

(c) The larva has a fixed number of cells.

(d) The larva is ciliated.

(e) The larva does not feed.

(f) On arrival in the host, the somatic cells degenerate and the generative cells develop.

(g) The larva is the distributive phase between one host and the next.

(5) The adult male Orthonectid has a reproductive duct.

Details of the structure and life history of the Mesozoa are given in Figs. 25–28. Our knowledge of the life cycle of the Orthonectids is fairly complete but that of the Dicyemids is not. Thus we do not know if they have a second host, and, if so, the morphology of the parasite in this host. A suggested life cycle is shown in Fig. 28 and this cycle is more complex than that described for the Orthonectids (McConnaughey 1951).

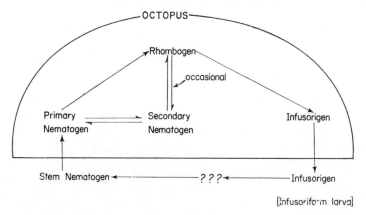

FIG. 28. Diagram of the life cycle of *Dicyema*. (After McConnaughey.)

Caullery (1951) stated that it was probable that the adult of Orthonectids such as *Rhopalura* would on future examination show greater histological differentiation. " A more careful histological analysis than so far made will probably disclose a nerve ring. What is lacking is a digestive apparatus, as in the Monstrillidae, and here this is almost certainly because the life of the adult is here even more ephemeral, and entirely devoted to the production and dissemination of larvae." Caullery is clearly of the opinion that the Orthonectids are degenerate forms.

There is no certainty that the Orthonectids and the Dicyemids are as closely related as their grouping together in the Mesozoa suggests. The resemblances are mainly that both have a ciliated larva and the adult structure is multicellular without a gut or organ systems. The life cycle of *Dicyema* is not yet fully known and so it is difficult to compare it with *Rhopalura*. There is a plasmodial stage in *Rhopalura* which has not been described for *Dicyema*. The ciliated larva is not identical in structure in the two forms.

If the Mesozoa are not primitive, they can be considered as degenerate digenetic trematodes, possibly the miracidium larva of some trematode that has become the end stage of development, e.g. the Mesozoa are neotenous miracidia. There are, as we have seen, certain resemblances between the miracidium and the ciliated mesozoan larva, but as yet no miracidium has been described which is as simple as the mesozoan larva. The mesozoan larva has a very short life and so it might not have time to develop the flame cells found in the miracidium. Perhaps some experimental studies on miracidia and the condition under which they can be maintained will allow us to come to a greater understanding of the Mesozoa. There is also a great deal to be discovered about the life history and habits of this group of animals before we can come to any conclusion about their phylogenetic position.

(3) THE COELENTERATA

Over the early years there was a controversy over the animal nature of the Coelenterata and it took some time before their true animal nature was recognised (Johnstone 1838). Even so, various groups of animals such as the Ectoprocta, Endoprocta and the Coelenterata were grouped together mainly on the similarities of their external form.

T. H. Huxley in 1849 presented a memoir on the anatomy and affinities of the Medusae to the Royal Society of London. In this memoir he described how the Medusae differed from the rest of the animal kingdom in that they could be regarded as having only two layers whilst the other metazoans had three layers. Huxley suggested that the layers in the Coelenterata were homologous with those of the Vertebrata and that in fact the Coelenterata were diploblastic.

In Huxley's text *A Manual of the Anatomy of the Invertebrated Animals* published in 1891 he modified his views a little. He classified the coelenterates into two main groups, the Hydrozoa and the Actinozoa, and he included the Medusae in the Hydrozoa.

(1) Hydrozoa: (*a*) Hydrophora *Tubularia*.
(*b*) Discophora *Aurelia*.
(*c*) Siphonophora *Physalia*.
(2) Actinozoa: (*a*) Coralligena *Actinia*.
(*b*) Ctenophora *Pleurobrachia*.

He compared the coelenterate body to a sac. " The walls of the sac are composed of two cellular membranes, the outer of which is termed the ectoderm, and the inner the endoderm, the former having the morphological value of the epidermis of the higher animals, and the latter that of the epithelium of the alimentary canal. Between these two layers, a third layer—the mesoderm—which represents the structures which lie between the epidermis and the epithelium in more complex animals, may be developed, and sometimes attains great thickness, but it is a secondary and, in the lower Hydrozoa, inconspicuous production. Notwithstanding the extreme variety of form exhibited by the Hydrozoa and the multiplicity and complexity of the organs which some of them possess, they never lose the traces of this primitive simplicity of organisation and it is but rarely that it is even disguised to any considerable extent. . . . In the fundamental composition of the body of an ectoderm and an endoderm, with a more or less largely developed mesoderm, and the abundance of thread cells, the Actinozoa agree with the Hydrozoa. . . . There is a certain similarity between the adult state of the lower animals and the embryonic conditions of the higher organisations. For it is well known that, in a very early state, even of the highest animals, it is a more or less complete sac, whose thin wall is divisible into two membranes, an inner and an outer. . . . There is a very real and genuine analogy between the adult Hydrozoön and the embryonic vertebrate animal, but I need hardly say it by no means justifies the assumption that the Hydrozoa are in any sense ' arrested developments ' of higher organisms."

From the above account by Huxley two points are clear. Firstly he thought that the resemblance between the embryonic development of the higher animals and the organisation of the coelenterates into two main layers of importance as indicating the primitiveness of the coelenterates. Secondly Huxley realised that mesoderm, or its precursor, did occur in the coelenterates and thought that there was an increase in the thickness and complexity of the mesoderm in the higher coelenterates. In effect he assumed that simple Hydrozoa such as *Hydra* and *Tubularia* were more primitive than the members of the Actinozoa.

There are now two questions that should be considered. The first is what are the simple and complex characters of the coelenterates? From a study of these it should be possible to assess how near the coelenterates are to the basic metazoans. The second problem concerns the relationship of the Hydrozoa,

Scyphozoa and the Actinozoa, and in effect revolves around which of these can be considered as being the most primitive. As we have seen, Huxley considered that the Hydrozoa were the more primitive, but many zoologists now think that the Actinozoa are the more primitive.

Let us now consider the simple and the complex characters of the coelenterates. Some of these, as shown in the following lists, are contradictory, but this is due to the wide range of structure occurring within the coelenterates.

Simple coelenterate characteristics

(1) There are only two well-developed epithelial layers, ectoderm and endoderm.
(2) They have a mesogloea.
(3) They have musculo-epithelial cells.
(4) The ectoderm may be ciliated.
(5) The gut has only one opening.
(6) There is a hydraulic skeleton.
(7) They are free living forms and not parasitic.
(8) Digestion is both intra- and extra-cellular.
(9) There is no respiratory or excretory system.
(10) They are polymorphic.
(11) They have a high regenerative capacity.
(12) They have a planula larva.
(13) The nerve net shows little concentration.
(14) They show radial symmetry.

Complex coelenterate characteristics

(1) They may develop cells in the mesogloea to form mesen-chyme and mesoderm.
(2) The body layers may become quite complex, e.g. three types of cells in the ectoderm: (*a*) sensory and mucus cells; (*b*) interstitial cells; (*c*) muscle cells.
(3) They may have separate muscle cells (Trachylina and Scyphozoa) which may be striated. The musculature can be complex, e.g. circular, longitudinal and oblique muscle bands.
(4) They develop a skeletal system. This may be an exoskeleton in *Obelia* or *Heliopora* or an endoskeleton as in *Corallium*.

(5) They have a gut. (The Mesozoa and Acoela have no gut.)
(6) The gut may have subdivisions (pharynx, mesenteries).
(7) The gut develops a circulation in *Aurelia* and *Alcyonium*.
(8) They have nematocysts.
(9) They have specialised sense organs such as eyes and statocysts.
(10) Some coelenterates are bilaterally symmetrical.
(11) Some coelenterates such as *Velella* and *Porpita* show division of labour.

As is well known, there is considerable diversity of structure within the coelenterates, and even though various coelenterates can be derived from a common plan there is still difficulty in deciding which are the most primitive coelenterates as opposed to the most simple. Thus though one can arrange a series going from, say, *Hydra* to *Physalia*, or from a diploblastic radially symmetrical form to one that is triploblastic and bilaterally symmetrical, there is no historical justification for either such series. We must find some collateral evidence to help determine which are the most primitive of the coelenterates.

The most primitive coelenterates

There is hardly a group of the coelenterates that has not at one time or another been claimed to have been the most primitive. Perhaps the two most prevalent claims are (1) that the polyp is the most primitive form, and (2) that the medusa is the most primitive form.

The view that the polyp is the most primitive form in the coelenterates has been supported by Haeckel, de Beer and Hadži as well as various other writers. Whilst Haeckel suggests that *Hydra* is primitive, Hadži thinks that the anemones are more primitive and that evolution within the coelenterates has gone from the Anthozoa to the Scyphozoa and Hydrozoa. This view is supported by de Beer (1954), who writes, " It follows and is generally recognised, that the polypoid person which is the only one represented in the Anthozoa, is more primitive than the medusoid person found in the Scyphozoa and Hydromedusae, which is clearly an adaptation to dispersal on the part of the sessile form."

It is interesting to compare the above statement with one taken from Hyman (1940). " The contrary theory, that the ancestral coelenterate was a primitive medusa, therefore seems more acceptable. This could readily have developed from the meta-gastraea by putting forth tentacles and when armed for food capture would not have been limited to a bottom habitat." R. C. Moore (1956) is of a similar opinion to Hyman. " Next the conclusion that the polypoid and medusoid types of organisation, instead of representing a more or less unexplained ' alternation of generations ' constitute the products of evolutionary differentia-tion in which the polypoid form is a persistent early growth, and the medusoid is the normal adult type of coelenterate, leads to the interpretation of medusoids as the initial type of coelenterate. This is consistent with the paleontological record, which includes numerous Lower Cambrian and even Precambrian medusoid fossils. Consequently the simplicity of the hydroid forms is not accepted as a basis for placing them in first position among various types of coelenterates. Precedence is assigned to early medusoids."

Though the medusa has been suggested as the basic form in the Coelenterata this has not been followed up by claiming that the Scyphozoa are the most primitive class of the Coelenterata. The life cycle of the Scyphozoa with their dominant medusa and their temporary polyp (hydratuba) might fit in with the primitive system. The Stauromedusae such as *Haliclystus* and *Lucernaria* indicate the way in which an adult polyp, even a highly specialised polyp, could have arisen. The nematocysts in the Scyphozoa are more limited in range of form than those in the Hydrozoa. It might be objected that the medusae of the Scyphozoa are very much more complex than those of the Hydrozoans, but the complexity of the present-day forms does not mean that the original forms were of the same complexity. The present-day forms and even the Cambrian fossils have a tremendous history of development behind them. The choice of a primitive class in the coelenterates will clearly depend upon the light that such a choice throws on our understanding of coelenterate morphology.

The most popular choice of primitive class in the coelenterates seems to lie between the Hydrozoa and the Anthozoa. Opinion is divided as to which of the Hydrozoa are the most primitive. Thus a selection of authors and their choice of primitive form is shown below.

Haeckel	Hydrida
Moser	Siphonophora
Hyman	Trachylina

It is difficult to choose between the above groups; thus though *Hydra* is more simple in its adult morphology, it is suggested from a study of the range of form that the medusoid condition is more primitive in the Hydrozoa and that development from this led to the solitary polyp.

On the other hand there is a growing body of opinion that the Anthozoa are more primitive than the Hydrozoa. This view is supported by Hadži (1944), Ulrich (1950), Remane (1955), Jagersten (1955) and Marcus (1958), who suggest the developmental sequence went Anthozoa–Scyphozoa–Hydrozoa. No closely reasoned account has yet been presented by the above authors to show exactly how the morphology of the primitive anthozoan would lead one to suppose that they are more primitive than the hydrozoans, but the gist of the evidence is apparently as follows.

(1) If the Hydrozoa were the most primitive forms which later gave rise to the Anthozoa this would not explain the marked bilateral symmetry found in the Anthozoa. Bilateral symmetry is usually associated with a mobile habit and one would not expect to find it in a sessile form that had a long sessile history behind it. This bilateral symmetry is found in the Ordovician Tetracorallia and even in the arrangement of flagella in the zoöxanthella larva. From the symmetry as shown in the arrangement of the mesenteries, the retractor muscles, septal filaments, siphonoglyphs and sulcus, one would suppose that the Anthozoa arose from a free-living mobile ancestor.

(2) A second reason for choosing the Anthozoa as the most primitive form lies in the range and structure of the nematocysts. The Hydrozoa have over a dozen different types of nematocysts whilst the Anthozoa have only about half a dozen different types. Furthermore the cnidoblast that carries the nematocysts is more simple in the Anthozoa; it lacks the cnidocil and instead has a primitive ciliary cone (Pantin 1942).

(3) The Anthozoa, Scyphozoa and many of the higher animals form their endoderm by invagination. This method is rarely found in the Hydrozoa, where ingression is more usual, and this latter situation has been regarded as being a specialised condition.

If we accept these reasons for choosing the Anthozoa as the primitive class of the coelenterates, what would the primitive form look like? It might have been something between an Antipatharian and a Protanthean. In the Antipatharia there are only six, ten or twelve septa. The longitudinal musculature on the septa is scanty or absent and the flagellated tracts are very simple. The mesogloea is scanty and the siphonoglyph only weakly developed. In *Protanthea* there are eight macrosepta and four microsepta. There is a complete cylinder of longitudinal epidermal muscles in the column and pharynx (these are much reduced in other Anthozoa). The nerve net and ganglion cells are well developed over the surface of the body—the ectodermal nerve net being reduced in other anemones. The sphincter and basilar muscles are absent. The retractor muscles are weakly developed and there are neither septal filaments nor a siphonoglyph.

Although forms such as *Antipathes* or *Protanthea* may be simple Anthozoa, they are still very complex when compared to a protozoan. We still know very little about the primitive anthozoans but it requires a lot of imagination to bridge the gap between the Antipatharia and the Protozoa.

Are the Coelenterata the most primitive of the lower Metazoa? We have to choose between the Mesozoa, Porifera, Coelenterata, Ctenophora and the Turbellaria to find the most primitive metazoan. It seems likely that the simplicity of the Mesozoa can be discounted as due to their entirely parasitic nature. Similarly the sponges can be discounted since their level of organisation is quite different in nature from that present in the other metazoan. It would be best to place the sponges on a side line to the main line of origin of the Metazoa; the time of origin of the side line is not clear.

There is no doubt that the simplest of the Hydrozoa are more simple than either the Ctenophora or the Turbellaria. But as we have already mentioned, we do not know that simple forms such as *Hydra* are the most primitive of the Coelenterata. One of the major clues that has been used to place the coelenterates has

been that of embryological development. Haeckel suggested that
the adult coelenterates such as *Hydra* were at a stage comparable
to the gastrula seen in *Amphioxus*. Even in Haeckel's time it was
pointed out that the embryology of the coelenterates did not
follow that of the higher animal. The blastula of the Hydrozoa
is most often a solid larva, the interior of which is filled with
cells. Hadži and de Beer take this solid larva to indicate that the
primitive larva had a solid gut and they think that the coelenterates
cannot be primitive since they have a hollow gut. It may be
correct that the most common type of planula has a solid
interior. But this in no way indicates that the adult also had a
solid gut.

The Acoela are not the only animals to have a solid gut in the
adult condition. Within recent years the Pogonophora, a group
related to the Pterobranchiate Protochordates, have been
described by Ivanov (1954–7). They have a solid gut filled with
endoderm cells. The Pogonophora are coelomate animals and it
is not clear whether the solid gut is here a primitive condition or
one that is due to the small size of the animal. At any rate it leads
one to wonder about the precise conditions that lead to the
retention of the solid gut if it is a primitive condition, or the
development of a solid gut if it is an advanced condition. Jagersten
(1955) thinks it highly unlikely that the primitive metazoans had a
solid gut since this would mean that the hollow gut arose at least
twice, once in the Coelenterata and again in the Turbellaria. It is
perhaps worth noting that certain coelenterates such as *Clytia*,
when they feed, fill the gastro-vascular cavity with endodermal
processes so that the gut takes on a solid mesh-like appearance.
Thus Hadži's view that the Coelenterata cannot be the most
primitive of the Metazoa since they have not a solid gut is open to
two objections: firstly we do not know that the solid gut is a
primitive condition and secondly some coelenterates can at times
show a condition resembling a solid gut.

In conclusion, then, it is apparent that we do not know whether
the coelenterates are more or less primitive than other lower
metazoans such as the Turbellaria. We do not know if the hollow
gut is a primitive condition. We do not know if the Hydrozoa are
more primitive than the Anthozoa. We do not know which is the
more primitive form, the medusa or the polyp, and as we shall

7—IOE

now see, we do not know the relationship between the Coelenterata and the Ctenophora.

(4) THE CTENOPHORA

There are three questions that should be discussed concerning the Ctenophora. (1) What is their ancestry? (2) Are they coelenterates? (3) Are they ancestral Turbellaria? These are all difficult questions to answer and involve a careful consideration of the structure of the ctenophores.

The ancestry of the Ctenophora

Nothing definite is known about the ancestry of the ctenophores. It is generally suggested that they arose from a basic stock that gave rise to the coelenterates; thus there are certain resemblances and certain differences between the ctenophores and the coelenterates, as can be seen from the lists below.

Coelenterate characteristics of the Ctenophora

(1) There are two primary layers; the ectoderm and endoderm are well developed and there is no definite mesoderm—just a mesenchyme.

(2) The main body cavity is the gastro-vascular cavity.

(3) There is only one opening to the gut—the mouth. There is no true anus.

(4) They have a stomodeum at the entrance to the gut as in the anemones and some medusae.

(5) The gut is divided like that of the Scyphozoa but the eight divisions are more like the symmetry of the Alcyonaria.

(6) They are radially symmetrical.

(7) They have mesenchyme muscles like some of the coelenterates, e.g. Trachylina, Scyphozoa.

(8) The gonads are derived from interstitial cells.

(9) The outer surface has cilia; as comb plates in *Pleurobrachia*, as a ciliated surface in *Coeloplana*.

(10) The tentacles are like those of some Scyphozoa.

(11) The lasso cells may take the place of nematocysts, but *Euchlora* has true nematocysts.

(12) There are no nephridia.

(13) They have a subepidermal nerve net.

(14) *Gastrodes* has a planula larva.
(15) Certain coelenterates such as *Hydroctena* resemble ctenophores.

Differences between Coelenterata and Ctenophora

(1) Some ctenophores have openings to the gut other than the mouth. (Similar openings are found in some medusae such as *Aequorea*.)
(2) The lasso cells are morphologically quite distinct from nematocysts. The nematocysts of *Euchlora* have been stated to have been derived from its food.
(3) There are no musculo-epithelial cells.
(4) *Coeloplana* and *Ctenoplana* have genital ducts.
(5) The cilia are arranged in specific rows, or comb plates (except for *Coeloplana*).
(6) Their symmetry is more biradial than radial.
(7) The embryology of the ctenophores is determinate and the cleavage differs markedly from that of the coelenterates.

Can the Ctenophora be placed in the Coelenterata? It can be seen that there are many resemblances between the coelenterates and the ctenophores. The two greatest differences seem to lie in the possession of nematocysts by the coelenterates and the embryology of the two groups. The ctenophores do possess lasso cells which differ in their morphology from the nematocysts found in the coelenterates. However, one ctenophore, *Euchlora rubra*, has been found to have nematocysts. When these were investigated by Komai (1942) he found that the nematocysts occurred in the tentacles and inside cells which he thought were endodermal cells. In 1951 Komai suggested that the nematocysts might have been taken from the food of *Euchlora* since the nematocysts did not lie on the surface of the tentacle but were sunk into the ectoderm. Hadži (1951) thought that the nematocysts in *Euchlora* were derived from its food, there being a close resemblance between its nematocysts and those of the narcomedusan *Cunina*.

Picard (1955) reinvestigated this problem and found that the nematocysts did in fact lie on the surface of the ectoderm, correctly orientated for discharge. The nematocysts were only found in association with ectodermal cells, never endodermal cells. Picard

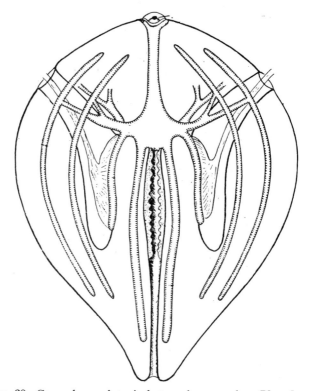

FIG. 29. Ctenophora. A typical ctenophoran such as *Pleurobrachia*
shown here is a round transparent animal with eight ciliated comb
rows. (From Hyman.)

suggested that the nematocysts were formed mainly during the
larval phase of the ctenophore. The nematocysts differ in
structure from those of the narcomedusae, which would indicate
that they cannot be derived from *Cunina*. Furthermore all the
specimens had nematocysts.

All these points lead one to conclude that *Euchlora* has its own
true nematocysts. This would then indicate that the ctenophores,
or at least this ctenophoran, belong to the Cnidaria!

The apparent wide embryological differences between the
coelenterates and the ctenophores may be diminished when we
know more about the range of embryological development of the

Scyphozoa. For the ctenophores in many ways resemble the scyphozoans; thus, both have a poor regenerative ability in the adult, they both develop thick mesenchyme and they both develop muscles. Considerable interest was aroused by the discovery of the medusa *Hydroctena*. Haeckel suggested that it was an ancestral form to the Ctenophora and it certainly shows a superficial resemblance to a ctenophore as can be seen from Fig. 31. The resemblances are due to the ovoid shape, the two tentacles (which Haeckel thought could be retracted into pockets at their base) and the gut being divided into four pockets. On the other hand there are many differences. *Hydroctena* has a circular canal around the perimeter of its body, the tentacles are oral and non-retractile (those of the ctenophores are aboral and retractile). There is no statocyst, it has nematocysts and not lasso cells, and the gonads develop on the wall of the manubrium instead of the radial canals. What is of interest here is the manner in which the ctenophoran form can be imitated by a medusoid form.

Another interesting coelenterate that shows certain ctenophore–turbellarian affinities is *Tetraplatia*. It was classified as a narcomedusan hydrozoan by Carlgren (1926) placed in a separate order of the Hydrozoa, the Pteromedusae by Hand (1955), but identified as a coronate scyphozoan by Krumbach (1927). Its external form is elongate and very much like that of a Müller's larva. There are eight lappets around the body, this giving some resemblance to a ctenophore; on the other hand the mouth is terminal (Fig. 31). *Tetraplatia's* affinities are further discussed by Ralph (1959).

It is unfortunate that the planula larva is not more common in the ctenophores. The only known case is in *Gastrodes*, which is parasitic on *Salpa*. Otherwise the ctenophores have a typical cydippe larva. There are some clear affinities between the coelenterates and the ctenophores but just how closely the two are related is hard to say. If further investigations show the presence of nematocysts to be more widespread than just in *Euchlora* and if they are of a similar pattern to the coelenterate nematocysts and not like those of the protozoan nematocysts, it will indicate that the ctenophores can be included in the Cnidaria. The relationship between the Hydrozoa, Scyphozoa and Anthozoa seems to be a closer one than that of these three to the Ctenophora.

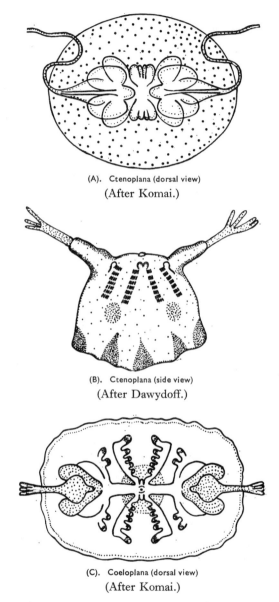

(A).   Ctenoplana (dorsal view)
(After Komai.)

(B).   Ctenoplana (side view)
(After Dawydoff.)

(C).   Coeloplana (dorsal view)
(After Komai.)

Fig. 30. Aberrant ctenophores. These ctenophores take up a crawling habit and their shape differs from that of *Pleurobrachia*.

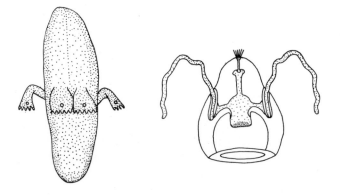

Tetraplatia   Hydroctena

FIG. 31. Aberrant coelenterates.

Tetraplatia shows superficial resemblance to the Müller's larva of
the polyclad platyhelminthes. (After Krumbach.)

Hydroctena. This medusoid form shows certain resemblances to
a ctenophore. (After Dawydoff.)

Relationship of the Ctenophora to the Turbellaria

Although typical ctenophorans such as *Pleurobrachia* or
Hormiphora are round pelagic animals there are some creeping
forms. It was these creeping forms and in particular *Coeloplana*
(Fig. 30) that led Lang (1884) to suggest that the ctenophores gave
rise to the polyclad Turbellaria. It is not hard to bridge the
gap between the pelagic forms such as *Pleurobrachia* and creeping
forms like *Coeloplana*. Thus *Lampetia* is a semi-globular form that
sometimes crawls on its everted pharynx. *Ctenoplana* in its
swimming form is clearly a ctenophore (Fig. 30B) but in its
crawling form it spreads itself out on its oral lobe and becomes a
flat animal. Finally *Coeloplana* is a flattened form like a turbellarian
(Komai 1922) and it looks very much like a link between the
ctenophores and the polyclad Turbellaria. In transverse section
Coeloplana has a complex structure (Fig. 33) and it is not surprising that Lang thought that it was the forerunner of the
turbellarians. In particular he associated it with the polyclads
because of the many branches of the gut. The polyclad resemblances can be seen from the following list.

Resemblances between *Coeloplana* and the Polycladida

(1) Both have a flat, compressed body.

(2) They move by creeping on the sole of the " foot."

(3) The body surface is ciliated.

(4) There is a well-developed basement membrane.

(5) The dermal musculature is well developed; the dorso–ventral muscles may be branched.

(6) The gastric canals have many branches; there is no anus but some pores end externally.

(7) There is a stomodeal invagination on the ventral surface.

(8) Both show determinate cleavage.

(9) The large micromeres give rise to small micromeres.

(10) The micromeres form the ectoderm.

(11) There is no hollow blastula stage.

(12) The development is mosaic.

(13) The mesoderm arises from the macromeres.

(14) The embryo gastrulates by epiboly.

(15) The Müller larva present in some polyclads has eight ciliated lappets.

(16) There is a small apical nervous tuft.

(17) There is a statolith.

(18) Both groups have paired tentacles.

(19) There are gonadial canals.

The above list is impressive in length and indicates a considerable similarity between the two groups. Lang suggested that the centrally positioned nerve centre in the ctenophores moved anteriorly to take up the typical polyclad position, i.e. a change from biradial symmetry to bilateral symmetry (Fig. 32). The list may also demonstrate another point. Often in discussing such a problem, a list of characters is drawn up of the points for and the points against a given viewpoint. The difficulty comes when one has to decide the relative importance of each character. It is impossible to come to a decision just by seeing whether there are, say, more similarities than differences. Instead each point must be weighted according to its importance and this is difficult since the importance often reflects the opinion of the observer.

In spite of the similarities between *Coeloplana* and the Polycladida, the general opinion these days is that the similarities are

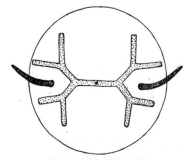

(A).   Ctenophoran condition

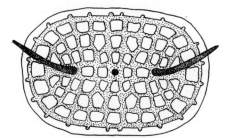

(B).   Hypothetical condition

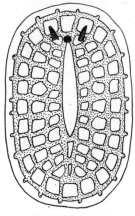

(C).   Polyclad condition

FIG. 32. Lang's concept of the manner in which the ctenophores
could have given rise to the polyclads. The anus moved ventrally
and backwards and the body form became elongated.

due to convergence and that *Coeloplana* and *Ctenoplana* are in fact
specialised aberrant and advanced forms. In particular there are
considerable differences between the embryology of the ctenophores and the polyclads which make it improbable that the two
groups are related. These differences are not due to the presence
or absence of yolk but instead reflect a more fundamental difference. The ctenophore egg cleaves into four and at the next
cleavage it divides to form a small group of cells, the micromeres,
and a large central group, the macromeres. These form a flat
plate of cells. Cleavage continues till there are eight macromeres
and many micromeres. In the polyclads, on the other hand, after
the four-cell stage the embryo shows a definite spiral cleavage
pattern. The cells can all be classified in terms of the spiral cleavage
pattern found in the Rhabdocoelida, Tricladida, Annelida and
Mollusca. This spiral cleavage is not found, nor is there any
indication of spiral cleavage, in the development of the
Ctenophora.

There are other differences between the polyclads and the
ctenophorans. Thus the polyclads have a well-developed brain,

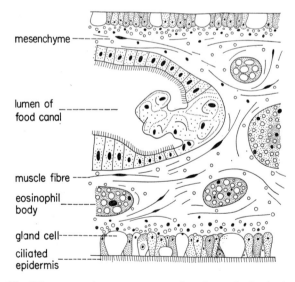

mesenchyme

lumen of
food canal

muscle fibre

eosinophil
body

gland cell

ciliated
epidermis

FIG. 33. Diagrammatic transverse section through the body of a
ctenophore, *Coeloplana*. (After Komai.)

often the most highly developed brain of all the turbellarians; they also have well-developed and numerous eyes, flame cells, a complex reproductive system with a muscular penis, uterus, seminal vesicle and prostate organ, and often show hypodermic impregnation. The Ctenophora have nothing to compare with this.

It is interesting to mention here that Hadži (1944) and de Beer (1954) think that the Ctenophora arose from the Polycladida, a viewed discussed in more detail on page 94. Hadži places the Ctenophora in the platyhelminthes though he agrees that *Coeloplana* and *Ctenoplana* are aberrant forms.

What conclusion can we come to regarding the position of the Ctenophora? With regard to their level of organisation they are in most respects at a similar level to that seen in the Anthozoa–Scyphozoa line of the Coelenterata. It is not possible to place them any closer than this until more research has been carried out on the embryology and development of the ctenophores and until the range of form of the coelenterates is better known. It is not

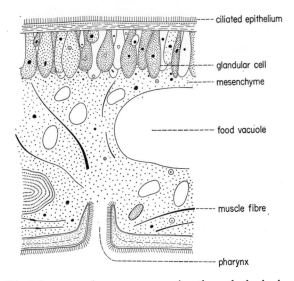

FIG. 34. Diagrammatic transverse section through the body of an Acoelan, *Convoluta*. There is a certain resemblance to the grade of organisation of the ctenophorans. (After von Graff.)

even certain at this stage that the ctenophores and coelenterates had a common origin. It is possible that the nematocysts of the ctenophores could have arisen independently of those in the coelenterates; after all there are some well-developed nematocysts in the Protozoa. On the other hand there is almost nothing to favour the view that the ctenophores are platyhelminthes. This is particularly so because the turbellarians have a well-developed reproductive system with accessory muscular sacs, whilst the most that any ctenophore has is a small reproductive duct. There is thus no clear indication that the ctenophores either gave rise to or were derived from the Turbellaria.

(5) THE PLATYHELMINTHES

Though the platyhelminthes are usually considered as having evolved after the Coelenterata, Hadži (1944, 1953) has suggested that this is not the case and that in fact the coelenterates evolved after and from the platyhelminthes. The classification that Hadži gives of the lower Metazoa is as follows.

Phylum	Subphylum	Class
Spongiaria	Spongiaea	
Ameria	(1) Platyhelminthes	Planuloidea
		Turbellaria
		Ctenophora
		Trematoda
		Cestoda
	(2) Cnidaria	Anthozoa
		Scyphozoa
		Hydrozoa

This classification differs from the usual one in several respects. First of all the platyhelminthes are considered to be the most primitive of all the Ameria, more primitive than the Cnidaria. Secondly the Ctenophora are placed in the platyhelminthes. Thirdly the Anthozoa are considered to be the most primitive of the Cnidaria. The evolutionary sequence devised by Hadži is as follows.

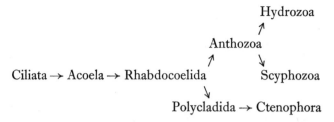

Hadži thinks that the Metazoa arose by the formation of cell walls in a syncytial ciliate. This would then lead to a multicellular animal with a complete organisation, i.e. an antero–posterior axis and without the difficulties of reorganisation that a multicellular

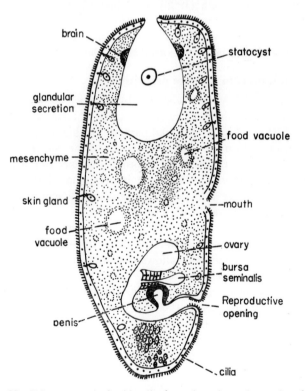

FIG. 35. Diagrammatic longitudinal section through an Acoelan to show the order of complexity of its structure. Note the development of a complex reproductive system. (From Bronn.)

colonial animal might have had. The syncytial level of organisation would correspond to the solid blastula, the stereoblastula,
that is sometimes found in the embryology of the Metazoa. The
simplest Metazoa did not have a hollow gut and corresponded in
structure to the present-day Acoela (Fig. 35).

The Acoela gave rise to the other platyhelminthes, amongst
which were the Rhabdocoelida with their straight gut, and the
Polycladida with their branched gut. The Rhabdocoelida gave
rise to the Anthozoa which in their turn gave rise to the Hydrozoa
and the Scyphozoa. The Polycladida have a swimming larval
form, the Müller's larva, and this became neotenous and gave
rise to the Ctenophora. Neotenous Müller's larvae have been
described; thus Heath (1928) described *Grafizoön lobata* which
resembled a sexually mature Müller's larva.

The arguments that Hadži puts forward in favour of his views
are as follows and he is supported by de Beer (1954, 1958).

The Coelenterata are not primitive

(1) The radial symmetry shown by the coelenterates is secondarily acquired by them. The development of bilateral symmetry is
shown first of all in the external parts of the higher animals and is
then impressed on the internal organs. In the coelenterates such
as the Anthozoa, the bilateral symmetry is only found internally
as in the mesenteries. Therefore it must have lost its external
bilateral symmetry and be in the process of acquiring a radial
symmetry concomitant with a sessile habit.

(2) The polyp is more primitive than the medusoid form.
Since the polyp is found in the Anthozoa whilst the Hydrozoa have
both polyp and medusa it follows that the Anthozoa are the most
primitive of the Coelenterata.

(3) The Coelenterata are not diploblastic. They have a well-
developed middle layer, the mesogloea, which often contains
cells. Furthermore the cell layers in the Anthozoa and the
Hydrozoa are not strictly comparable since in the Hydrozoa the
germ cells are formed from the ectoderm whilst in the Anthozoa
and the Scyphozoa they are formed from the endoderm. The
germ layers in the coelenterates are not comparable to the germ
layers of the higher animal and are not even homologous within
the coelenterates.

(4) Haeckel suggested that the Coelenterata represented the hollow gastrula stage found in the embryology of the Echinoderms, *Sagitta* and *Amphioxus*. This would indicate that the coelenterates are a primitive group. But we now know that the blastula is more often a solid form and that the endoderm is not always formed by the invagination and development of a hollow gut. In fact the hollow gut is an advanced condition when compared with that present in the Acoela.

Objections can be raised to all of Hadži's points.

(1) There is no evidence that the radial symmetry is secondarily acquired by the coelenterates. The detection of an external symmetry depends upon having some external organs; the coelenterates do not have any such organs along the length of the polyp and thus they cannot display this external bilateral symmetry. It cannot therefore be established that they had an external bilateral symmetry at some stage which was later lost. Furthermore there is no evidence that bilateral symmetry is acquired first of all by the external organs and later by the internal organs.

(2) It is not generally accepted that the polyp is the most primitive form in the coelenterates. In fact there is quite a body of opinion that holds the medusa to be the primitive form (see p. 80).

(3) Though it is correct that not all the coelenterates are diploblastic, the mesogloea of many of the Hydrozoa shows little or no development. They at least can be considered as having an effective diploblastic condition. As for homologising the cell layers, it is not strictly correct to assert that the gonads arise from the ectoderm in the Hydrozoa. It is more correct to state that they arise from interstitial cells. In this way, therefore, one can homologise the germ layers within the coelenterates.

(4) Finally the fact that the blastula may often be solid in no way indicates that the adult must have had a solid gut in the most primitive metazoans. The larval form merely indicates what the primitive larval condition was like, not the adult condition.

Having asserted the non-primitive nature of the coelenterates Hadži presents the following evidence that the Acoela are more primitive than the coelenterates.

The Acoela are primitive

(1) The Acoela organisation corresponds to that of a ciliate that has developed cell walls.

(2) The Acoela have no gut.

(3) The Acoela are hermaphrodite and have internal fertilisation rather like the syngamy of the ciliates.

(4) The Acoela often have a syncytial gonad, epidermis, reproductive system and digestive system. This is a relic of the original ciliate syncytium.

(5) The digestive system of the Acoela can be derived from the protozoan food vacuoles (Figs. 5 and 35).

(6) The nephridial system can be derived from the protozoan contractile vacuoles. The Acoela do not have flame cells.

(7) The Acoela usually have a ciliated ectoderm.

(8) The Acoela often have musculo-epithelial cells.

(9) The Acoela have no basement membrane.

(10) The Acoela have a central mouth like the ciliates.

(11) The Acoela have a simple pharynx derived from a stomodeum.

(12) There are no distinct gonads.

(13) The rhabdites are derived from the trichocysts.

The Acoela and the Polycladida differ in their morphology and development from the rest of the Turbellaria. Hadži suggests that the Acoela gave rise to the Rhabdocoelida and the Polycladida. The Rhabdocoelida then gave rise to the Anthozoa by the loss of their protonephridia, the reduction of their nervous system, the simplification of their digestive system, the loss of accessory reproductive organs and reduction of the mesoderm. The slime glands of the epidermis of the rhabdocoels gave rise to the nematocysts.

The derivation of the Anthozoa from the Rhabdocoelida in this way would be very surprising with no other parallel in the animal kingdom, i.e. reduction and simplification giving rise to a whole phylum of widely diverse and successful animals. This does not mean that such a reduction is impossible; it just seems highly improbable. It also seems unlikely that the polyclads gave rise to the ctenophores.

There is also little assurance that the Acoela are the most primitive of the Turbellaria. They are a comparatively unstudied group of animals; we know next to nothing about their physiology, little experimental embryology has been performed on them and we do not know their range of morphological forms. Though von Graff (1904) was of the opinion that their structure was primitive, a great deal more research will have to be carried out before such a position can be justified and even more research will be necessary before they can seriously be derived from the ciliates. For further details concerning the phylogeny of the platyhelminthes and their relationship to the Ctenophora one should consult Bresslau (1933).

There is a school of thought represented by Marcus (1958) which suggests that the platyhelminthes are an advanced group of animals that were once coelomate and more complex morphologically than, say, the Nermertini, Phoronidea or Brachiopoda. The platyhelminthes according to this view are secondarily simplified. They have lost their coelom, anus and circulatory system. They have reduced their nervous system, and altered their reproductive system to make up for the loss of the coelom. It is suggested that one well-known example of a coelom being lost when an animal takes up parasitic habit is that of the Hirudinea, and that the platyhelminthes have gone even farther along this course.

We thus have two conflicting views concerning the status of the platyhelminthes. Hadži suggests that they are the most primitive of all the Metazoa, being more primitive than the coelenterates and derived from the ciliates. Marcus suggests that the platyhelminthes are an advanced group of animals whose simplicity of structure is due to their parasitic habit and that they arose some time after the Nemertea.

What can one conclude about the most primitive of the Metazoa? There are, as we have seen, five contestants, Porifera, Mesozoa, Coelenterata, Ctenophora and the Platyhelminthia, for this title. These groups are almost completely isolated from each other though a few tenuous connexions can be made. It is quite clear that the available evidence is insufficient to allow us to come to any satisfactory conclusion regarding their interrelationships. At the same time it is also clear that a great deal of work remains to be done on all of these groups. We are still very ignorant about the comparative physiology and biochemistry of the lower Metazoa and very little experimental work has been done on their embryology. It is possible that these lines of research will help in the elucidation of the relationships between the lower Metazoa. It is also possible that the new information will indicate more clearly that the Metazoa are polyphyletic.

CHAPTER 7

THE INVERTEBRATE PHYLA

WITHIN the invertebrates there are many distinct phyla. So far we have considered some of the possible relationships between the so-called " lower phyla," namely the Protozoa, Porifera, Mesozoa, Coelenterata, Ctenophora and Platyhelminthia. There are, however, many other important phyla such as the Nematoda, Nemertea, Rotifera, Annelida, Arthropoda, Mollusca, Brachiopoda, Echinodermata and Protochordata that all deserve some mention for they each present special problems of phylogenetic relationship.

It is not possible to obtain satisfactory palaeontological *data* concerning the relationship of these various phyla because most of them are already fully established in the earliest fossil-bearing beds, the Cambrian. This means that one has to use other information to determine the relationships between these phyla. In fact these relationships are not at all clear and this can best be illustrated by examining three attempts that have been made to present a coherent monophyletic relationship of the major invertebrate phyla.

GROBBEN'S CLASSIFICATION

Karl Grobben in 1908 proposed a scheme to show the interrelationship of various invertebrate groups. This system has formed the basis for most of our current schemes, e.g. Cuénot 1952. The system divides the major invertebrate phyla into two sections, the Protostomia and the Deuterostomia. This distinction had been proposed by Goette in 1902 and was based on the fate of the blastopore in the developing embryo: whether it becomes the anus or the mouth and anus. In the Protostomia the blastopore becomes the mouth and anus whilst in the Deuterostomia it

101

becomes the anus; the mouth develops in another position. Grobben then divided the phyla in the following fashion.

Protostomia	*Deuterostomia*
Scolecida	Chaetognatha
Molluscoidea	Echinodermata
Mollusca	Enteropneusta
Annelida	Tunicata
Arthropoda	Acrania
	Vertebrata

By Scolecida, Grobben meant the Platyhelminthia, Entoprocta, Aschelminthia and Nemertini. In the term Molluscoidea he included the Phoronidea, Ectoprocta and Brachiopoda. Sometimes the group Molluscoidea was referred to as the Tentaculata.

Let us first of all consider the validity of these two major groups, the Protostomia and the Deuterostomia. In the Protostomia the situation is not as clear cut as the classification might suggest. Thus in the platyhelminthes, the blastopore closes in *Convoluta* and *Planocera* and the mouth is a new formation (there is of course no anus in the platyhelminthes). In the polyclads the original blastopore closes and disappears but the pharynx develops near the site of the erstwhile blastopore. In the Tardigrada the blastopore does not develop. In the Entoprocta the blastopore closes and a new mouth and anus develop. In the Annelida the situation varies according to the animal studied. In *Nereis* and *Podarke* the blastopore forms the mouth and anus in the required manner. In *Pomatoceros* the blastopore forms the mouth but the anus is a new formation. In *Capitella*, *Ctenodrilus* and *Saccocirrus* the blastopore closes and the mouth and anus are new formations. In the oligochaete *Dendrobaena* the anus is a new formation and it is not derived from the blastopore.

In the Arthropoda the situation is much the same. In *Peripatus capensis* the blastopore after a brief closure opens again to form the mouth and anus. In some crustaceans such as *Caridina* the anus develops some distance away from the blastopore region whilst the mouth develops as a new formation unrelated to the blastopore. In *Astacus* the proctodeum arises from the region near the site of the blastopore.

In the Mollusca the fate of the blastopore also varies. In the Gastropoda as a rule the anterior part of the blastopore gives rise to the mouth, in *Paludina* the mouth arises from the posterior part of the blastopore. In all gastropods the anus is a new formation unrelated to the blastopore. In the Amphineuran *Ischnochiton*, too, the anus is a new formation. The lamellibranches such as *Teredo* or *Cyclas* have the blastopore closed completely and the mouth and anus are entirely new formations. Otherwise the blastopore becomes the site of the mouth. Further information can be found in Dawydoff (1928) and Manton (1948).

In the Deuterostomia the fate of the blastopore is similarly varied. The echinoderm mouth is a new formation in the larva whilst the blastopore becomes the larval anus. The same is true for the hemichordates. In the Tunicata and the Cephalochordata the blastopore becomes dorsally placed to form the neuropore. In the Chaetognatha the blastopore closes and does not become either the mouth or the anus.

It can be seen that the division into Protostomia and Deuterostomia is not as sharp as might be expected. Thus certain groups such as the Tardigrada, Chaetognatha, Tunicata, Cephalochordata and so on would be in neither the Protostomia nor the Deuterostomia. In other groups such as the Annelida or Arthropoda, certain genera have the blastopore forming the mouth whilst others do not.

This situation was appreciated quite early. Thus Sedgwick stated in 1915, " In *Peripatus* the mouth and anus are not only derived from the elongated blastopore by its constriction into two openings but remain throughout life included within the nerve ring derived from the neural rudiments of the embryo. If in other Arthropoda, in Annelida, and in the Mollusca we find, as we do, that the nerve ring referred to is, in the adult, incomplete behind the anus, and the mouth and anus, though obviously referable to the blastopore, are not actually derived from it, must we on this account deny this most obvious relation and maintain that the mouth or anus, as the case may be, in these forms is not homologous with that of *Peripatus*? To maintain such a position appears to us impossible and we entirely accept the doctrine that the mouth and anus of the Annelida, Arthropoda, and Mollusca are both perforations of the embryonic neural surface and are

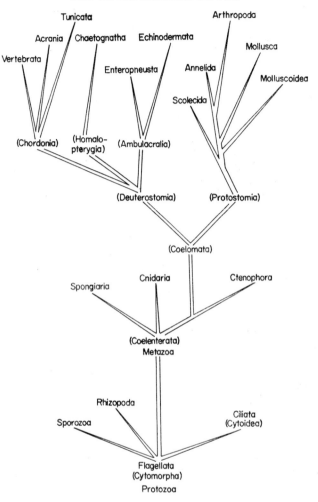

FIG. 36. Grobben's classification of the Invertebrate Phyla.

specialisations of parts of one original opening which is represented in most embryos by the blastopore.

"When, however, we come to apply this doctrine to the Chordata we stand on more debatable ground. Placing the Enteropneusta on one side as not obviously conforming to our plan, we find that it is a fact of observation that in the Chordata the blastopore perforates the embryonic rudiment and that in some of them the anus is directly derived from it. (Many Pisces, some Amphibia, e.g. Newt.) Whereas in others not at all remote from these, the blastopore closes entirely and the anus is a new formation (some Pisces and Amphibia, e.g. Frog, Amniota). Here also we think it may be fairly maintained that notwithstanding the diversity in the mode of development of the anus, it is, in all vertebrata at least, a derivative of the blastopore."

It will be seen from Sedgwick's account that whilst he realised there was considerable diversity in the fate of the blastopore, he thought the generalisation—that the blastopore became the mouth in the Protostomia and the anus in the Deuterostomia—a fair one. On the other hand Manton (1948) thinks that the variability in the mode of development of the mouth and anus is so great in the Annelida and the Arthropoda that it no longer forms a useful link between these two groups. " It is clear that most of the known species of Onychophora fall into line with the Arthropoda in the dissociation of the mouth and anus from the blastoporal area and contrast with the majority of the Polychaeta."

There are other characters that can be used to separate the Protostomia and the Deuterostomia, or the Annelid and the Echinoderm Superphylum as they are sometimes called.

Annelid Superphylum	*Echinoderm Superphylum*
Spiral cleavage	Radial cleavage
Blastopore = mouth	Blastopore = anus
Schizocoelic coelom	Enterocoelic coelom
Determinate cleavage	Indeterminate cleavage
Nervous system delaminates	Nervous system invaginates
Ectodermal skeleton	Mesodermal skeleton
Trochosphere larva	Pluteus type larva

To each and every one of these characters many exceptions can be found and in particular, certain groups of animals seem to lie between the two superphyla. Thus the Brachiopoda have their blastopore forming the mouth, their coelom is enterocoelic and their cleavage is of the radial type. The situation is most difficult for the Nematoda, Ectoprocta and Phorodinea and these are the groups that one would most like to place accurately. Even within the major groups there is some disagreement. Thus Raven in his account of morphogenesis in the Mollusca states that it is incorrect that the group as a whole shows determinate cleavage. This would indicate that the Mollusca–Annelida link is not necessarily as close as some authors imagine. On the other hand it is necessary to keep some sense of balance and not lose sight of the wood because of the trees.

Grobben's classification is shown in Fig. 36. In some ways it resembles the next classification to be discussed, that of Marcus, but there are certain differences. Thus the Coelomata arise from the line that led to the Ctenophora. The Enteropneusta are more allied to the Echinodermata than they are to the Tunicata or Vertebrata.

MARCUS'S CLASSIFICATION

This view of the phylogeny of the invertebrates has been described by Marcus (1958) and it agrees in many ways with that described by Grobben and also with that described by Ulrich (1950) and Remane (1954). Marcus considers that the Anthozoa are the most primitive of the Coelenterata. All the forms above the Coelenterata are called " Bilateria " since they are almost all bilaterally symmetrical. They are also called " Coelomata " and Marcus considers that all these forms are derived from an ancestor that had the "fundamental features of the Archicoelomata, viz. three coeloms, mouth, anus, vessels and perhaps tentacles." The coelom was developed as a series of pouches from the gut, as suggested by Sedgwick (1884), and the Bilateria could have arisen from either the Anthozoa or the Ctenophora.

Since the Bilateria are all coelomate this means that the Platyhelminthia, Rotifera, Nematoda and Endoprocta all are derived from a form that once had a coelom. During the course of evolution the coelom became reduced in these forms till sometimes all that is left is the cavity of the flame cells. The resemblance that has been reported between the planula larva and the Acoela, Marcus suggests, is entirely due to the small size of the animals.

The Bilateria are divided by Marcus into the Protostomia and Deuterostomia, as we have already seen in Grobben's classification, the only addition being the newly described Pogonophora, which are placed in the Deuterostomia. Marcus points out that there are certain real resemblances between the animals at the base of the Protostomia and the base of the Deuterostomia. Thus the Ectoprocta and the Pterobranchiata have a body that is in three segments; the Ectoprocta and Brachiopoda are enterocoelic, the Ectoprocta have coelomic pores and budding is often similar in pattern.

The first major division within the Protostomia is the Tentaculata, which is comparable to the Molluscoidea of Grobben. Of these the Phoronidea are considered as being the most primitive whilst the Entoprocta are considered to be derived from attached larvae of the Ectoprocta.

The Nemertea are coelomate, their coelom being restricted to the rhynchocoel and the gonad cavities. The nemerteans are more primitive than the Turbellaria since they have an anus and their

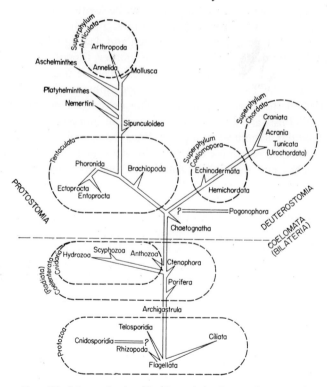

FIG. 37. Marcus's classification of the Invertebrate Phyla.

genitalia are more simple. Their embryology indicates a possible relationship with the polyclads.

The Platyhelminthia have most of their organ systems reduced and simplified; the animals are not therefore primitive. The coelom is reduced to the ciliated ducts of the reproductive organs. The Acoela are not the most primitive of the Turbellaria (Fig. 37).

The Aschelminthia include the Nematoda, Rotatoria, Gastrotricha, Nematomorpha, Kinorhyncha and Priapulida, some of which show a spiral pattern of cleavage. It is not clear if the Aschelminthia are a closely related group of animals.

The Mollusca and Annelida are derived from a common ancestry. The ventral pharyngeal sac of the archiannelids is similar to the radula sac of the molluscs and the teeth of the Eunicidae show plates that are similar to the radula teeth. The primitive molluscs such as *Neopilina* may be segmented.

The Articulata (Arthropoda) arose several times from the annelid stock. The Pentastomida, Onychophora and Tardigrada are three groups that are quite distinct from one another, though similarities between the legs, body cavity and gonads can be used to form a link between the Tardigrada and the Onychophora. The Trilobita gave rise to the Arachnomorpha. The crustacean resemblances of the trilobites are due to homoiology; the independent derivation of similar structures in separate lines that are phylogenetically related. (Other examples of homoiologous organs are compound eyes, trachea and malpighian tubules.) The basic line that gave rise to the Crustacea also gave rise to the Antennata from which came the Myriapoda and the Insecta.

The Deuterostomia are a smaller and more compact group than the Protostomia. The hemichordates contain the Enteropneusta and the Pterobranchiata. The Enchinodermata and the Enteropneusta are linked together by the dipleurula larva. The ancestor of the Hemichordata then being postulated as giving rise to the Tunicata and the Vertebrata.

Hadži Classification

The relationship that Hadži (1944, 1957) postulates between the various invertebrate groups can be seen from Fig. 38. He derives the Metazoa from the Ciliophora. In the ciliates there is

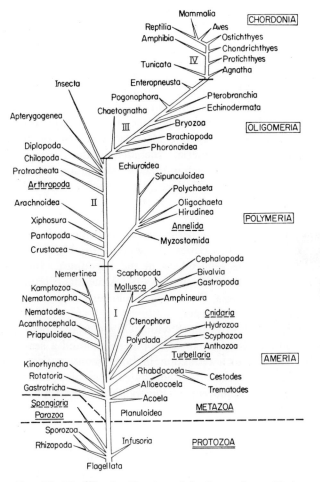

FIG. 38. Hadži's classification of the Invertebrate Phyla.

often a differentiation of the cytoplasm in a manner that can be compared with the ectoderm, mesoderm and endoderm of the metazoa. Hadži thinks that such a ciliate gave rise to a form resembling an acoelous turbellarian and that the Turbellaria are the most primitive of the Metazoa. The Turbellaria then gave rise to the Anthozoa, as has been mentioned on p. 95.

The coelom of the metazoa is traced back to the mesohyal

(mesoderm) of the ciliates; it is not therefore a new formation. Various cavities such as the nephridial cavity, blood cavity, secretory cavities, lymph cavities, rhynchocoel and pericardium are all regarded as being part of the coleomic system. There are primary cavities, those without an epithelial lining, and secondary cavities, those with an epithelial lining. The perigastrocoel is a space lying alongside the gut and it too has become lined with epithelium. Those animals that have such a perigastrocoel and which are also unsegmented Hadži places in his first metazoan phylum, the Phylum Ameria. Included in the Ameria are the following groups: Platyhelminthia, Coelenterata, Gastrotricha, Rotatoria, Kinorhyncha, Mollusca, Priapuloidea, Acanthocephala, Nematoda, Nematomorpha, Nemertini and Kamptozoa.

The second phylum is the Phylum Polymeria. These animals are all segmented and the perigastrocoel becomes initially broken up to form a series of cavities. In some of the higher Polymeria the cavities become reduced, as in the Hirudinea. Included in the Polymeria are the following groups: Annelida, Sipunculoidea, Echiuroidea, Crustacea, Pantapoda, Xiphosura, Arachnida, Chilopoda, Diplopoda and Insecta.

The third phylum is the Phylum Oligomeria. These animals have at some stage of their development adopted a sessile habit and this has led to a reduction in body segmentation. In the Oligomeria are placed the Phoronidea, Brachiopoda, Bryozoa, Chaetognatha, Echinodermata, Pogonophora, Enteropneusta and Pterobranchiata.

The fourth phylum is the Phylum Chordonia. In this are placed the Vertebrata and the Tunicata.

Hadži does not consider that the higher invertebrates can be satisfactorily classified into Protostomia and Deuterostomia. Instead he thinks that the line that gave rise to the higher arthropods also gave rise to the echinoderms.

Marcus's classification and that of Grobben have more in common than either has to Hadži's, the greatest difference being in the position of the platyhelminthes; Grobben thinks they are primitive, Marcus thinks they are advanced. The major difficulty in their classification lies in the placing of phyla other than the Annelida, Mollusca, Echinodermata and Protochordata. All the remaining small phyla are difficult to place. It is possible that some of them had an independent evolution from the Protozoa. Even within the major groups such as the Arthropoda difficulties arise; it is becoming more certain that the Arthropoda are not a monophyletic phylum of animals but instead are a grade of organisation and that this grade has been reached independently many times from some annelid-like stock (Tiegs and Manton 1958).

It would appear that the relationship between the various invertebrate phyla is a very tenuous one. There are many phyla that seem to be isolated from each other, and even those phyla that seem reasonably close to one another, on detailed examination show differences as important as their similarities. Though it is useful to consider that the relationships determined by comparative anatomy and embryology give proof of a monophyletic origin of the major phyla, this can only be done by leaving out much of the available information. Let us now consider the invertebrate relationships determined by comparative biochemistry and see if they lead to any more definite conclusions.

CHAPTER 8

BIOCHEMICAL STUDIES OF
PHYLOGENY

THE PREVIOUS discussions concerning the phylogeny of animals has been concerned with evidence based mainly on morphological *data*. Within recent years, however, biochemical studies have been used to help determine animal relationships and the results so obtained have aroused considerable interest. Only two such studies will be considered here: those concerned with the distribution of phosphagens and those concerned with the distribution of sterols through the animal kingdom.

Due mainly to the work of the Cambridge biochemists, considerable interest has been focused on the phosphagens present in the invertebrates and the work has led to the discovery of a series of new and interesting chemical compounds. It is now intended to discuss the " phosphagen story " in some detail.

(1) PHOSPHAGENS

In many of the textbooks on comparative biochemistry or physiology such as those of Baldwin (1940) or Prosser (1952) one will find the following table.

Phylum and Class	Arginine phosphate	Creatine phosphate
Platyhelminthia	+	−
Annelida	+	−
Arthropoda	+	−
Mollusca:		
Lamellibranchiata	+	−
Cephalopoda	+	−

Echinodermata:
  Asteroidea $+$ $-$
  Holothuroidea $+$ $-$
  Echinoidea $+$ $+$
Protochordata:
  Tunicata $+$ $-$
  Enteropneusta $+$ $+$
  Cephalochorda $-$ $+$
Vertebrata $-$ $+$

This table indicates that most of the invertebrates have one type of phosphagen (arginine phosphate) whilst the vertebrates have another (creatine phosphate). The echinoderms and the protochordates have both types of phosphagen and this makes it seem likely that they are the group of invertebrates most closely related to the vertebrates.

Let us now consider the situation in a little more detail.

In 1927 Eggleton and Eggleton showed that one could extract a labile organic phosphorus-containing compound from vertebrate muscle. This compound was called phosphagen and later workers showed that it was in fact creatine phosphate. After isolation creatine phosphate broke down to form creatine and phosphoric acid.

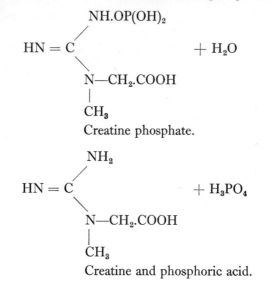

$$NH.OP(OH)_2$$
$$HN = C$$
$$N—CH_2.COOH$$
$$CH_3$$
$$+ H_2O$$

Creatine phosphate.

$$NH_2$$
$$HN = C$$
$$N—CH_2.COOH$$
$$CH_3$$
$$+ H_3PO_4$$

Creatine and phosphoric acid.

Creatine phosphate (CP) was present in many vertebrate muscles, thus the Eggletons found it in the muscles of *Amphioxus*, dogfish, plaice, frog, snake, tortoise, rabbit and guinea-pig. They were unable to find it in any of the invertebrate muscle they studied (*Aurelia, Lumbricus, Aplysia, Pecten, Holothuria*). Meyerhof (1928) found that there was a phosphagen present in invertebrate muscles but that it was not creatine phosphate but arginine phosphate instead. This he found in *Sipunculus, Pecten, Holothuria* and *Stichopus* muscle.

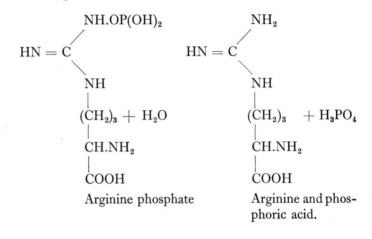

Arginine phosphate   Arginine and phosphoric acid.

Although it is not strictly within the scope of this book it might be as well to indicate the function of the phosphagens. They act as an energy reserve for muscle contraction. The phosphagens are high-energy compounds and they can phosphorylate adenosine diphosphate (ADP) to form adenosine triphosphate (ATP).

$$ADP + CP = ATP + C$$

When a muscle contracts and performs work, at some stage of the contraction–relaxation cycle it uses up the ATP and converts it to ADP. This ADP is reconverted to ATP by means of the phosphagen. At a later stage, glycolysis (the breakdown of glucose to carbon dioxide and water) brings about the synthesis of more high-energy compounds and the phosphagen is re-formed. This reaction $CP + ADP = C + ATP$ is sometimes called the Lohmann reaction and the reader can find more details in most texts on biochemistry (Harper 1959; Baldwin 1957; Fruton and Simmons 1958).

A thorough survey of the distribution of phosphagens CP and AP in the animal kingdom was published in 1931 by Needham, Needham, Baldwin and Yudkin. In some cases they dissected out the muscle tissue from the lower animals; in other cases they used the whole animal, the method used depending upon the size and availability of the raw material.

It should be remembered in all the discussions of their experimental work that most of the workers were pioneers in the field and that present-day criticism of techniques is in no way meant to be disparaging. It is only too easy to look back over a quarter of a century of research and, being wise after the event, to point out the various faults and errors. It is inevitable in a scientific subject that the years will bring great improvements in techniques which will then indicate that the previously used methods and conclusions were not sufficiently justified. There is but one way of making sure that one's work will never contain any errors and that is to do no work.

The technique that Needham *et al.* used for their analysis of the phosphagen was as follows. They cooled their material and dissected out the required part. This was then weighed, ground up with trichloracetic acid, left for 10 min in the cold and then filtered. The filtrate was neutralised with NaOH and then $CaCl_2$ was added to precipitate the inorganic phosphate. This precipitate of insoluble calcium phosphate was spun down in a centrifuge and separated from the supernatant fluid. The precipitate was dissolved in a few drops of concentrated sulphuric acid and the inorganic phosphate then determined.

The organic phosphate was still in the supernatant solution and it might contain the two possible phosphagens, creatine phosphate and arginine phosphate (CP and AP). These were analysed as follows. If one places CP (or AP) in acid solution, it hydrolyses to form either creatine (or arginine) and phosphoric acid. If molybdate ions are present the CP hydrolyses much more rapidly than does AP. Thus the determination of phosphate after 15 min hydrolysis gave an indication of the CP value whilst estimation after 15 hr gave both CP and AP values. The value of AP could then be determined by subtraction.

The results from Needham *et al.*'s experiments are often summarised as in the table on page 112, but I should like to present them here in slightly more detail. (See also the table on p. 117.)

(a) Coelenterates

In *Anthea rustica* 5·07 g of tentacles gave 0·053 mg of total phosphate of which 0·04 mg was due to inorganic phosphate. An experiment on 1·05 g of body wall gave a total phosphate of 0·032 mg and an inorganic level of 0·032 mg. A further experiment on 2·71 g of tentacle from *Anthea cereus* gave total phosphate of 0·135 mg and an inorganic level of 0·135 mg. The general conclusion from these experiments was that the level of phosphagens in anemones was too low to be detected.

In the ctenophore *Pleurobrachia pileus*, 33·67 g of total body gave a total phosphate of 0·12 mg whilst the inorganic phosphate came to 0·069 mg. This gave 42% organic phosphate which could be due to phosphagen.

(b) Platyhelminthes

0·4 g of *Planaria vitta* gave 0·056 mg of total phosphate of which 0·042 mg was due to inorganic phosphate. This gave a value of 24% AP. 0·54 g of *Polycelis nigra* had a total phosphate value of 0·098 mg and an inorganic phosphate value of 0·084 mg. Hence the value of AP was 14%.

(c) Nemertines

2·1 g of the whole body of *Lineus longissimus* gave a total phosphate of 0·987 mg whilst the inorganic phosphate came to 0·47 mg. This gave a value of 52% for AP. A second reading taking 1·45 g of body gave a total phosphate of 0·299 mg and an inorganic phosphate of 0·245 mg. The AP value came to 18%.

(d) Annelids

Three annelids were analysed, *Nereis*, *Sabellaria* and *Spirographis*. Two experiments were carried out on *Sabellaria alveolata*. In one case 1·38 g of whole body were taken which gave a total phosphate of 0·505 mg, 24% of which was due to phosphagen. In the other case 2·45 g of body gave a value of 0·789 mg of total phosphate of which 30% was due to phosphagen.

In *Spirographis brevispira* 2·83 g of the body gave 0·768 mg of total phosphate of which 63% was due to phosphagen. Another estimation from 3·08 g of body gave 0·925 mg of total phosphate of which 67% was due to phosphagen.

TABLE 1. SELECTED FROM INFORMATION IN NEEDHAM, NEEDHAM, BALDWIN AND YUDKIN (1931) SHOWING THE AMOUNTS OF INORGANIC AND ORGANIC PHOSPHORUS-CONTAINING COMPOUNDS IN VARIOUS INVERTEBRATES.

GROUP	ANIMAL	WEIGHT OF TISSUE	TOTAL P	INOR-GANIC P	PHOS-PHAGEN AS % OF TOTAL P
Coelenterata	*Anthea rustica*				
	tentacle	5·07	0·053	0·04	?
	body wall	1·05	0·032	0·032	0
	Anthea cereus				
	tentacle	2·71	0·135	0·135	
Ctenophora	*Pleurobrachia pileus*	33·67	0·120	0·069	42·0
Platyhelminthes	*Planaria vitta*	0·40	0·056	0·042	24·8
	Polycelis nigra	0·54	0·098	0·084	14·8
Nemertina	*Lineus longissimus*	2·10	0·978	0·470	52·5
Annelida	*Sabellaria alveolata*	1·38	0·505	0·383	24·1
	Spirographis brevispira	2·83	0·768	0·281	63·5
	Nereis diversicolor	3·45	1·359	0·781	50·3
Sipunculoidea	*Sipunculus nudus*	1·46	0·660	0·195	71·0
Cephalopoda	*Sepia officinalis*				
	fin muscle	1·25	0·935	0·79	15·3
	mantle	1·19	2·73	2·50	8·3
	Octopus vulgaris	1·66	1·77	1·42	12·7
Echinodermata	*Cucumaria planci*	0·49	0·037	0·037	0·0
	Synapta inhoerens	0·90	0·425	0·315	25·9
	Strongylocentrotus lividus	1·63	0·247	0·020	92·0
	Asterias glacialis	3·84	0·308	0·072	76·2
Protochordata	*Balanoglossus salmoneus*	0·36	0·101	0·057	42·8
	Ascidia mentula	7·03	0·05	0·039	22·5

In *Nereis diversicolor* seven normal animals were taken in which the phosphagen ranged from 15% to 81% of the total phosphate. It is of interest that in these measurements the authors found that whilst for the other annelids mentioned the phosphagen was always AP, in *Nereis* there appeared to be quite a lot of CP in five out of the seven normal samples. Thus out of the 15–81% due to total phosphagen the amount due to CP was 5–57%. The authors considered that these values of CP were due to errors in their technique and that creatine phosphate was not actually present.

In *Sipunculus nudus* one measurement was made from 1·46 g of body wall. This gave a total phosphate level of 0·66 mg of which 71% was due to AP.

(e) Cephalopoda

In *Sepia officinalis* various parts of the animals were analysed for phosphagen with the following results:

	Expt. 1	Expt. 2	Expt. 3
Fin muscle	4%	15%	7%
Mantle	13%	8%	0%
Funnel	6%	10%	0%
Tentacle	12%	5%	0%

In *Octopus vulgaris* the values from one animal came to, mantle 33%, funnel 27% and tentacle 12% phosphagen. It is of interest that in both *Octopus* and *Sepia* some of the phosphagen was apparently due to CP. In five out of the nine cases where phosphagen was present in *Sepia* there were traces of CP present, in one case it being one-third of the total 6% due to phosphagen. In *Octopus* in one case out of the three CP was present, it making up half of the total 12% due to phosphagen.

(f) Echinoderms

Two measurements on the body wall of *Cucumaria planci* showed no phosphagens to be present. However, two other experiments on the phosphagens of the body wall of another species of holothurium, *Synapta inhoerens*, gave values of 5% and 25% of the total phosphate content as being due to phosphagen. The phosphagen was AP. Nine experiments on the jaw muscles from the Aristotle's lantern of *Strongylocentrotus lividus* gave values from 42%–92% of the total phosphate as being due to phosphagen. Analysis showed that about one-third or more of this phosphagen was due to CP.

In the spines and muscles of *Echinocardium cordatum* it was not possible to detect any phosphagens at all.

In the tube feet of the starfish *Asterias glacialis* there was only AP present; this made up 76% and 73% of the total phosphate.

(g) Protochordates

In *Balanoglossus salmoneus* three readings were taken from different parts of the body. These showed a range in values of phosphagen of 14%, 16% and 42% of the total phosphate. In the case where 42% of the phosphate was due to phosphagen it was shown that 21% was due to CP whilst the other 21% was due to AP. In the 14% of phosphagen the value was all due to CP whilst the other case of 16% was all due to AP.

The values given for *Ascidia mentula* were given as 22% and 12% of the total phosphate being due to phosphagen (AP). But as the authors pointed out, from 7 g of tissue they obtained only 0·05 mg of total phosphate and 0·039 mg of inorganic phosphate, hence the results were not very reliable.

On page 289 of their paper, Needham *et al.* drew up the following table.

The animal kingdom could be subdivided into animals that had

    (1) *No arginine phosphate:*
        Coelenterata   *Anthea*

    (2) *Only AP:*
        Coelenterata   *Pleurobrachia*
        Platyhelminthia   *Planaria, Polycelis*
        Nemertini   *Lineus*
        Annelida   *Sabellaria, Spiro-graphis, Nereis*
        Cephalopoda   *Sepia, Octopus*
        Echinodermata   *Synapta, Asterias*
        Urochorda   *Ascidia*

    (3) *CP and AP:*
        Echinodermata   *Strongylocentrotus*
        Hemichorda   *Balanoglossus*

    (4) *CP only:*
        Cephalochorda   *Amphioxus*
        Craniata   Many species

They concluded, " If any evolutionary significance may be attached to these findings, it is probable that they support the Echinoderm–Enteropneust (*Balanoglossus*) theory of vertebrate descent rather than any of the other views which from time to time have been put forward on this question."

In fact the situation was not quite so simple. Thus though the majority of invertebrates have arginine phosphate and the vertebrates have creatine phosphate, the authors found creatine phosphate in the Annelida (*Nereis*) and in the Mollusca (*Sepia*) as well as in the echinodermata (*Strongylocentrotus*) and Protochordata (*Balanoglossus*). In *Balanoglossus* it was present in only two of the three specimens analysed.

It would seem that there are no very good grounds for concluding from the phosphagen evidence alone that the echinoderms and the protochordates are more closely related to the vertebrates than are, say, the Annelida or the Mollusca.

In 1936 Baldwin and Needham repeated some of the determinations of phosphagens in the echinoderms. There were two problems in which they were interested; the first concerned the formation of the phosphagen. There should be an enzyme present in the tissue that would bring about the phosphorylation of the nitrogenous base and Baldwin and Needham decided to investigate the properties of this enzyme. Secondly they were not sure about the nature of the nitrogenous base in the phosphagen. Though they had felt it might be arginine and/or creatine their tests for these compounds were not specific tests but general ones. Thus the Sakaguchi test for arginine (make the solution alkaline with NaOH; add a little α-naphthol, then add a drop of sodium hypochlorite solution—a bright red colour develops) is not really specific for arginine but is given by the radical marked by a ring.

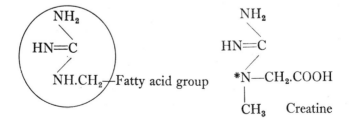

It will not react with creatine since creatine has a CH_3 group substituted for the H on the N marked with a *. However, other compounds such as glycocyamine (this is creatine without the CH_3 group and with an H instead) will give a positive reaction.

$$NH_2$$
$$\diagdown$$
$$HN{=}C$$
$$\diagup$$
$$NH.CH_2.COOH$$

Glycocyamine

The creatine was determined by the Jaffe reaction, which again is not an absolutely specific test for creatine.

The arginine from the phosphagen in the echinoderm muscle was thus tested by the Sakaguchi test, by seeing if the muscle extract could synthesise a phosphagen from arginine and phosphate, and thirdly by adding arginase and seeing if urea was given off.

The jaw muscles from *Sphaerechinus granularis* were shown to be capable of synthesising AP (26% of the added phosphoric acid being converted) and CP (12% of the added phosphoric acid being converted).

Extracts from the longitudinal muscles of *Holothuria tubulosa* gave extracts that could synthesise AP from A and P (though it should be noted that the Russian workers Verbinskaya, Borsuk and Kreps (1935) found AP and CP in the muscles of the holothurian *Cucumaria frondosa*).

Needham and Baldwin also quoted work done by P. Baldwin on the ophiuorid *Ophioderma longicauda* and the crinoid *Antedon mediterranea*. *Ophioderma* had CP whilst *Antedon* had AP.

The conclusion the authors drew from their work was that the ophiuroids and echinoids and possibly the holothurians had AP and CP and the other echinoderms (asteroids, crinoids) had only AP.

Let us see to what use this information is put. Hyman (1950) in her text on the echinoderms states on p. 700, " Further biochemical evidence supporting the close relationship of echinoids and ophiuroids concerns phosphagens, or phosphorus carriers, of

great importance in metabolic processes. . . . Crinoids, holothurians and asteroids have arginine as the phosphorus carrier
whereas creatine serves this function in the ophiuroids and the
echinoids (echinoids also have phosphoarginine). Phospho-
creatine is also characteristic of vertebrates; creatine in organisms
results from the methylation of glycocyamine and only echinoids
and ophiuroids have the enzymes (methylases) necessary for
performing this reaction. The author is of the opinion that the
closer relationship of ophiuroids to echinoids rather than to
asteroids, as usually supposed, is not to be doubted and therefore
the union of asteroids and ophiuroids into one group is not
admissible. Further the arrangement recently adopted by
palaeontologists according to which the asteroids and ophiuroids
derive from a common somasteroid ancestor and hence are to
be united into one class Stellasteroidea must somehow be wrong."

As the reader will have noticed in the analysis of Needham *et al.*
(1932) of the phosphagens in the annelids, they found some CP
as well as AP present in *Nereis*. They decided that this was possibly
due to some fault in their technique and that really only the AP
was present.

The problem was reinvestigated by Baldwin and Yudkin in
1949. They used similar techniques to the previous ones; thus
they differentiated between CP and AP by the rate of hydrolysis
in molybdate solution and they also used the Sakaguchi and the
Vosges–Proskauer test for the amino-acids arginine and creatine,
though neither of these tests, as they pointed out, were absolutely
specific.

Twenty-four different species of polychaetes were tested and in
addition they examined *Lumbricus*, *Phascolosoma* and *Sipunculus*.

The polychaetes are listed below.

Amphitrite johnstoni	*Lepidometria commensalis*
Amphitrite ornata	*Lumbrinereis sp.*
Arabella iricolor	*Maldane urceolata*
Arenicola marina	*Neanthes virens*
Branchiomma vesiculosum	*Nereis cultrifera*
Chaetopterus variopedatus	*Nereis diversicolor*
Cirratulus grandis	*Orbinia ornata*

Cistenides gouldii	*Pista palmata*
Clymenella torquata	*Sabella pavonia*
Diopatra cuprea	*Sabellaria alveolata*
Enoplobranchus sanguinea	*Spirographis brevispira*
Glycera dibranchiata	*Sthenelais leidyi*

They found that many of these animals had CP and AP. Thus *Amphitrite, Arenicola, Cirratulus, Clymenella, Enoplobranchus, Maldane, Nereis cultrifera, Pista, Sabella, Sabellaria* and *Spirographis* had AP but no CP. In *Chaetopterus, Diopatra, Glycera, Lumbrinereis* and *Orbinia* there was CP but no AP. In *Lumbricus* there was neither CP nor AP. *Phascolosoma* and *Sipunculus* had only AP. The other polychaetes had AP and CP.

The first impression of Baldwin and Yudkin (1948) was that there was a correlation between the occurrence of AP and the sedentary habit since *Amphitrite, Sabella, Sabellaria* and *Spirographis* all had AP and were sedentary, but further investigation (1949) showed that there was no such correlation. Thus even closely related genera had different phosphagens: *Neanthes virens* (which used to be called *Nereis virens*) had both CP and AP whilst *Nereis diversicolor* had only AP.

There was some doubt whether CP was really creatine phosphate and AP, arginine phosphate. Thus though CP on hydrolysis gave a positive Vosges–Proskauer test the authors concluded only provisionally that it was creatine phosphate and designated it as ' CP ' and not CP.

Similarly the arginine phosphate gave a very weak Sakaguchi reaction and they doubted if 'AP ' was arginine phosphate. They preferred to refer to it as annelid phosphagen—' AP.'

The fact that annelids have CP is of importance in deciding the phylogenetic importance and significance of the phosphagen. Thus previously it was shown that the invertebrates had AP whilst the vertebrates, some echinoderms and some protochordates had CP. We now see that polychaetes (twelve out of the twenty-four tested) had CP. This means that either the presence of CP is not a very good phylogenetic indicator or else that the annelids are more closely related to the echinoderms and the vertebrates than they are to, say, the molluscs.

Baldwin and Yudkin (1949) also carried out some analyses of the phosphagens in echinoderms and protochordates. They concluded that the hemichordates and the echinoids were unique in that they both had CP and AP whilst the other echinoderms (except for ophiuroids) had only AP. These results were presented in tabular form and indicated that the vertebrates were derived from the Echinoderm–Protochordate line.

In fact the evidence for the phylogenetic value of phosphagens is not very good. Out of the three hemichordates studied, *Balanoglossus salmoneus*, *Saccoglossus kowalewsky* and *Saccoglossus horsti*, only the former has AP whilst the others have CP. Rees (1958) states that in his analysis of twenty specimens of *Balanoglossus clavigerus* he was able to find only CP; there was no indication of AP.

Amongst the echinoids the jaw muscles of *Arbacia punctulata* have only AP and no CP whilst *Strongylocentrotus lividus* and *Echinus esculentus* have both CP and AP. Similarly Griffiths, Morrison and Ennor (1957) showed that though some echinoids such as *Heliocidaris erythrogramma* had both CP and AP, others such as *Centrostephanus rodgersi* had AP and no CP. These authors concluded, " The general assumption of Baldwin and Needham that both AP and CP are found in the echinoids is thus disproved and the results emphasise the necessity for examining a number of species within a class before concluding that a particular phosphagen is characteristic of the class." If one was to take the possession of the phosphagens as a serious phylogenetic feature one might conclude that since the ophiuroids have only CP like the vertebrates that they in fact are the closest of the echinoderms to the vertebrates.

The phosphagen story took a new turn in the 1950s when chromatographic analysis was applied to the guanidine compounds. As we have already seen, the previous workers were only concerned with two guanidines, creatine and arginine, and they differentiated these on the rate of hydrolysis and various non-specific tests. French workers at the Laboratory of Comparative Biochemistry of the College de France (van Thoai, Roche, Robin and Thiem, 1953) showed that the annelids contained at least two other phosphagens. They found these by running ascending chromatograms of muscle extracts from *Arenicola marina* and *Nereis diversicolor* in either pyridine–water or propanol–acetic–water and developing the chromatograms in α-naphthol hypobromite which gives a coloured spot with guanidines. They found that in *Arenicola* there was the compound taurocyamine phosphate whilst in *Nereis* there was glycocyamine phosphate.

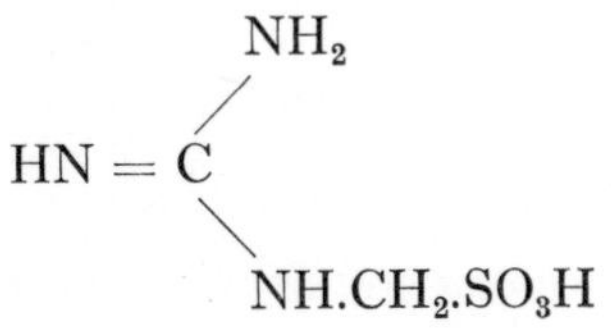

$$HN = C \begin{cases} NH_2 \\ NH.CH_2.SO_3H \end{cases}$$

Taurocyamine

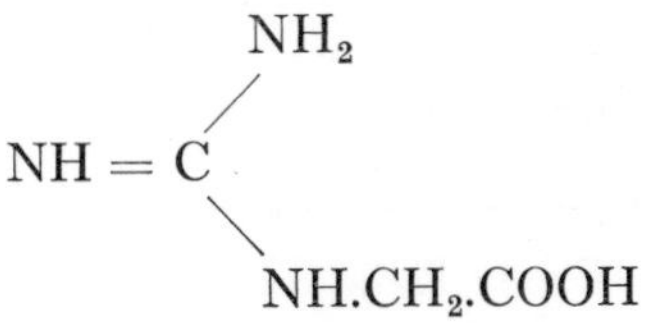

$$NH = C \begin{cases} NH_2 \\ NH.CH_2.COOH \end{cases}$$

Glycocyamine

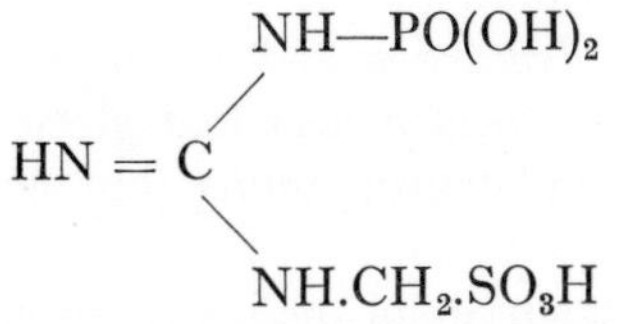

$$HN = C \begin{cases} NH—PO(OH)_2 \\ NH.CH_2.SO_3H \end{cases}$$

Taurocyamine phosphate

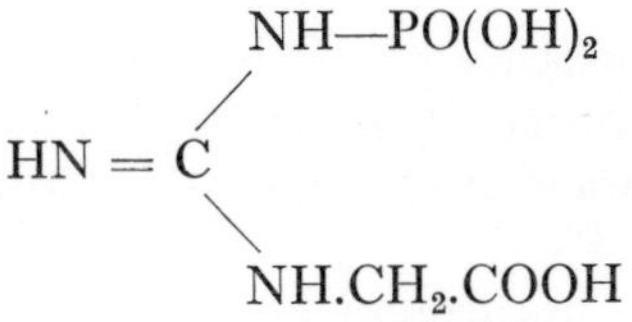

$$HN = C \begin{cases} NH—PO(OH)_2 \\ NH.CH_2.COOH \end{cases}$$

Glycocyamine phosphate

The relationship of these compounds to the other phosphagens was not as obscure as might be supposed on first sight. Thus van Thoai and Robin (1951) had shown that an enzyme capable of methylating various compounds had quite a wide distribution in the invertebrates and that it was quite possible that this might methylate glycocyamine to form creatine.

It is probable that glycocyamine phosphate (GP) and taurocyamine phosphate (TP) play a similar role in the body to AP and CP since there are enzymes that can phosphorylate G and T. Thus Hobson and Rees (1957) showed that specific phosphokinases were present in various annelids. The unphosphorylated base was added to the muscle extract, inorganic phosphate and the appropriate buffer. This was then incubated at 40 °C for 15 min and the phosphagens formed isolated and tested. The results are shown in the following table.

TABLE 2

ANIMALS	μM OF PHOSPHATE FORMED			
	TP	GP	AP	CP
Arenicola marina	5·0	1·25	0	0
Nereis diversicolor	0·4	6·5	0	0
Nereis fucata	0	5·8	0	2·0
Hermione hystrix	0	0	0	4·5
Aphrodite aculeata	0	0	0	1·25
Myxicola infundibulum	1·5	0	0	0
Nephthys cacea	0	1·7	0	0·7

From Table 2 it can be seen that *Arenicola* and *Myxicola* have TP whilst *Nereis diversicolor* has GP as its main phosphagen. *Hermione* and *Aphrodite* have CP whilst none of them had AP.

About the same time Roche and Robin (1954) showed the presence of CP in the sponge *Thetia lyncurium* and AP in the sponge *Hymeniacidon caruncula*. *Hymeniacidon* also had glycocyamine (but not GP) whilst *Thetia* had taurocyamine (but no TP).

Roche *et al.* (1957) made a thorough survey of the distribution of arginine and creatine in the animal kingdom, using the methods of chromatographic separation and semi-specific chemical tests. The species that they analysed and the results they obtained are shown below.

Species	Creatine	Arginine
Sphaerechinus granularis	+	+
Martasterias glacialis	+	+
Amphipholis squamata	+	+
Ophiothrix fragilis	+	+
Leptosynapta inhoerens	+	+
Maia squinado	0	+
Apis mellifica	0	+
Bombyx mori	0	+
Sepia officinalis	0	+
Helix pomatia	0	+
Limnaea stagnalis	0	+
Mytilus edulis	0	+
Ostrea edulis	0	+

Species	Creatine	Arginine
Arenicola marina	+	+
Audouinia tentaculata	+	+
Clymene lumbricoides	+	+
Dasybranchus caducus	+	+
Glycera convoluta	+	+
Lineus marinus	0	+
Lumbriconereis	+	+
Marphysa sanguinea	+	+
Nephthys hombergi	0	+
Nereis diversicolor	0	+
Sabella pavonina	0	+
Scolophus armiger	+	+
Lumbricus terrestris	0	+
Hirudo medicinalis	0	+
Phascolosoma elongatum	0	+
Sipunculus nudus	+	+
Ascaris lumbricoides	0	+
Actinia equina	0	+
Anemonia sulcata	+	+
Calliactis parasitica	+	+
Halichondria panicea	+	+
Hymeniacidon caruncula	+	+
Thetia lyncurium	+	+
Tetrahymena geleii	0	+

The authors conclude that the very wide distribution of creatine does not allow one to come to any conclusion concerning its phylogenetic importance and in particular the presence of creatine in the echinoderms in no way indicates an affinity or relationship with the vertebrates. This view is supported by Ennor and Morrison (1958) in their review of the biochemistry of phosphagens and related guanidines.

What conclusion can be drawn with regard to the distribution of phosphagens in the animal kingdom? The first conclusion is that there is certainly no simple cleavage of the animal kingdom into vertebrates with CP and invertebrates with AP. Instead it is clear that both CP and AP are found throughout the invertebrates. The second conclusion is that one cannot base any phylogenetic speculation on the occurrence of CP or AP since related genera within a class can differ widely in their phosphagens. The third conclusion is that it is highly probable that other phosphagens in addition to the recently discovered GP and TP will be found to play a role in tissue metabolism.

Thus Seaman (1952) has isolated a phosphagen from *Tetrahymena gelei* that does not appear to be any obvious guanidine derivative, i.e. not arginine, creatine, taurocyamine or glycocyamine. Van Thoai and Robin (1954) isolated another phosphagen from the muscles of *Lumbricus* and called it lombricine. Its structure is shown below. Lombricine has been further analysed by Beatty *et al.* (1959), who have shown that it contains a D-amino-acid, D-serine.

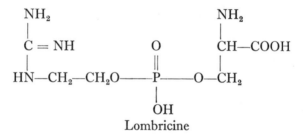

Lombricine

Robin, van Thoai and Pradel (1957) described a new guanidine derivative in the leech *Hirudo medicinalis* but have not as yet published details of its structural formula.

The function of a scientific theory is to help in our understanding of various pieces of information and to suggest further experiments that will test the validity of the theory. The value of a theory lies in the extent to which it stimulates the development of such new experiments. The theory concerning the distribution of phosphagens throughout the animal kingdom as suggested by Baldwin and Needham has been a very valuable one when judged in this manner.

It is becoming clear that the initial impetus that Baldwin, Needham *et al.* gave to the study of phosphagens has gathered momentum and a great deal of new information has been gathered concerning the chemical nature of phosphagens. What is now required is a very thorough analysis of material from many genera throughout the whole of the invertebrates for information as to the variety of phosphagens. It is certain that many new chemicals still remain to be discovered and the biochemical variations will probably be found to be as great as the more obvious morphological variations.

From the phylogenetic point of view, therefore, the phosphagens are not a great deal of assistance. The table shown on p. 112 can now be amended as below.

Phylum	AP	CP	Other guanidines
Protozoa	+		+
Sponges	+	+	+
Coelenterata	+	+	+
Platyhelminthia	+		
Nemertina	+		
Nematoda	+		
Annelida	+	+	+
Arthropoda	+		
Mollusca	+	+	
Echinodermata	+	+	
Protochordata	+	+	
Vertebrata		+	

The gaps in the table will be filled as more research is done on this subject.

(2) STEROLS

The phosphagens are not the only compounds that have been used to indicate phylogenetic relationships. Within recent years certain sterols have been used to elucidate relationships, though the work is still at a developmental stage. Some applications of the sterol studies are as follows.

Hyman (1955) in her volume on the echinoderms does not follow the normal custom and place the asteroids with the ophiuroids; instead she places her chapters in the order Holothuria, Asteroidea, Echinoidea, Ophiuroidea. Her reasons for doing this are stated on p. 699. In particular she decides that the ophiuroids and the asteroids should be separated and the ophiuroids placed with the echinoids on the basis of larval development, the possession of an epineural canal in the ophiuroids, and the possible occurrence

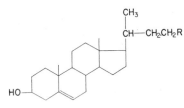

Cholesterol

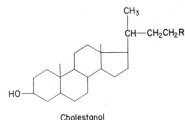

Cholestanol

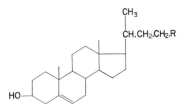

Clionasterol

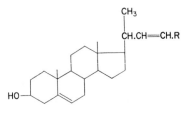

Porifasterol

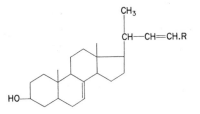

Stellasterol

FIG. 39. Sterol structure. This figure shows the structure of various of the sterols mentioned in the text.

R $= -CH_2.CH(CH_3)_2$ for Cholesterol and Cholestanol.

R $= -CH.(CH_3).CH.(CH_3)_2$ for the others.

of a vestibule in the ophiuroids. On p. 700 she states, " Finally, in recent years workers in comparative biochemistry have produced striking evidence in favour of this community of ancestry " (i.e. the ophiuroids with the echinoids and not with the asteroids). " Bergman (1949 and in a letter) finds that all ophiuroids and echinoids tested have sterols of Type I, namely cholesterol or some closely related compound, whereas numerous asteroids tested have Type II sterols that is, stellasterol or related compounds. The sterols of the three crinoids thus far tested belong to Type I, although perhaps a new variety, and those of holothurians classify as Type II. Further biochemical evidence supporting the close relationship of echinoids and ophiuroids concerns phosphagens or phosphorus carriers, of great importance in metabolic processes."

Bergman (1949) gives the following table showing the various species of echinoderms studied and the sterols present in each (see Fig. 39).

Asteroidea	Stellasterol	Hitodestrol	Cholesterol
Asterias rubens	*		
Asterias forbesi	*		
Asterias rollestoni		*	
Asterias scoparius		*	
Asterias pectinifera		*	
Echinoidea			
Tripneustes esculentus			*
Centrechinus antillarum			*
Lytechinus variegatus			*
Heliociderus crassidus			*
Arbacia punctulata			*
Holothuria			
Holothuria princepo	*		
Cucumaria chronjhelmi	*		
Ophiuroidea			
Ophiopholis aculeata			*?

From this table one can see that the asteroids and the holothurians both possess stellasterol whilst the echinoids and possibly the ophiuroids have cholesterol. This would link the asteroids and the holothurians on the one hand and the echinoids and ophiuroids on the other hand, an arrangement which would agree with that based on larval characteristics.

Perhaps it will pay us to look at the steroid situation in a little more detail. Bergman (1949) has given an interesting review of the distribution of lipids in marine invertebrates with special reference to the sterols. At one time it was thought that cholesterol was the only sterol present in these bodies but later work showed that

10—IOE

there were in fact over twenty different sterols, including optical isomers, present in the invertebrates and that it is quite likely as research proceeds that still more will be discovered. These sterols differ in (1) the length of the side chain, (2) the presence or absence of double bonds in this side chain, (3) the location of double bonds in the main sterol skeleton.

Up to 1949 the most studied phylum was the Porifera. This was due to the fact that they were easily obtained in fairly large quantities. Over fifty different species of sponges have been analysed by Bergman and his colleagues and they have obtained some very interesting results. In the first place they have discovered more than ten different sterols in sponges, only two of which had been known before. Secondly the presence or absence of these sterols helped in the elucidation of certain systematic problems.

For some time a sponge from the Biscayne Bay, Florida, had been given a variety of names. Some collectors had called it *Suberites distortus*, others *Suberites tuberculosus*. Bergman (1949) studied the sterols present in this sponge and showed that clionasterol and poriferasterol were present. Now these sterols were normally not found in the family Suberitidea but instead were more often found in the Clionidae or the Choanatidea.

	Cholest- anol	Clionast- erol	Porifer- asterol	Neospongo- sterol
Choanitidae		*	*	
Suberitidae	*			*
Clionidae		*	*	
Suberites distortus		*	*	

Suberites was very carefully examined by Laubenfells, who showed that there were some small microscleres present in the tissues. These microscleres were diagnostic of the genus *Anthosigmelia*, which is in the family Choanitidea. This then would mean that the sponge was not a Suberitidae but a Choanitidae and this would agree with the sterol assay.

A more complex case is present in that of *Hymeniacidon heliophila*. This sponge has been described both as *Hymeniacidon heliophila* and as *Stylotella heliophila*. The genus *Stylotella* is normally placed in the family Suberitidae, a family containing only saturated sterols. Bergman has shown that saturated sterols are present in this sponge and therefore the sterol assay agrees with its placing amongst the Suberitidae. Unfortunately the range of sterols of the family Hymeniacidonidae is not yet known but there is no reason why this sponge should be transferred from the genus *Hymeniacidon* to that of *Stylotella*. Bergman states, " such a transfer will remain premature until more is known about the sterol contents of other species of *Hymeniacidon* and of the closely related *Halichondria* " and he makes his position even more clear in the statement, " It is dangerous and frequently misleading to base significant conclusions concerning comparative biochemistry on *data* derived from but a few representatives of a given phylum. This point is apparent when the great diversity of sterols in Porifera is considered."

To return to the echinoderms it will be remembered that so far only three types of sterol have been described in the literature for the echinoderms. There is a likelihood that further study of the echinoderms will show a greater divergence of the sterols present in this group and that the cleavage shown in the table on p. 131 will not be so clearly delineated.

CHAPTER 9

VERTEBRATE PALAEONTOLOGY

THE MOST important evidence for the theory of Evolution is that obtained from the study of palaeontology. Though the study of other branches of zoology such as Comparative Anatomy or Embryology might lead one to suspect that animals are all inter-related, it was the discovery of various fossils and their correct placing in relative strata and age that provided the main factual basis for the modern view of Evolution.

It is unfortunate that the earliest rocks to contain fossils, the Precambrian and Cambrian, already show representatives of all the major invertebrate phyla. The earliest rocks are mainly igneous and it is possible that the fossils that they once contained have since been boiled away, but there is an alternative view that the invertebrates suddenly and explosively evolved and had little or no Precambrian history. Though there is some development of the various invertebrate fossils, especially within the phyla, our main examples of the evolution of the major groups of animals come from our study of the vertebrates. If we ask an under-graduate to give a brief account of the way in which the vertebrate palaeontology provides evidence for evolution, his answer may go rather like this.

" It is possible to date the rocks fairly accurately and in general the oldest rocks are at the bottom and the youngest rocks are on the top. There are sometimes cases where the rocks have been turned over so that the layers are sideways on or upside down, but careful study soon indicates this and allows one to determine their correct relative positions. If one studies the vertebrate remains, one finds that there are no vertebrate fossils in the oldest rocks. The next oldest rocks have some vertebrate fossils; these are fragments of simple fishes. The next oldest rocks have fish and amphibian fossils, the next have fish, amphibian and reptile fossils, whilst the most recent rocks will have fish, amphibian, reptile and mammal fossils (see Fig. 40)."

"The most important point is that one never finds a mammal fossil in rocks that are pre-reptilian; in fact the finding of a single mammal fossil in such an early stratum would seriously question the correctness of evolutionary concepts. Such a fossil has never been found and the evidence now accumulating strongly supports the view that the fish gave rise to the amphibia, the amphibia to the reptiles, and the reptiles to the mammals."

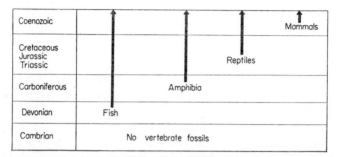
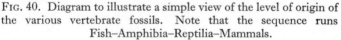

Coenozoic				Mammals
Cretaceous Jurassic Triassic			Reptiles	
Carboniferous		Amphibia		
Devonian	Fish			
Cambrian	No vertebrate fossils			

FIG. 40. Diagram to illustrate a simple view of the level of origin of the various vertebrate fossils. Note that the sequence runs Fish–Amphibia–Reptilia–Mammals.

This account, though a simple one, contains one serious fault. The figure shows not the time of *origin* of the different classes of the vertebrates but instead the time of *dominance* of that class. If we consider the time of origin we get a more complex picture (Fig. 41). Thus instead of having the reptiles, amphibia, bony fish and elasmobranch fishes all separated from each other by hundreds of millions of years, they all arose during the course of less than 100 million years. It is of course difficult to decide just when any of the groups did arise, but some estimate can be made.

The earliest fossil vertebrates, the Agnatha, are found in the Silurian (fragments are found in the Ordovician). The next group, the Placoderms, are found in the Upper Silurian. The bony fish arose in the Devonian as did also the elasmobranches and the Amphibia. (It is of interest to note here that there is one school of thought, examplified by Save Soderberg (1934) and Jarvik

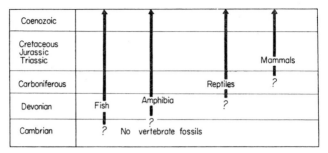

FIG. 41. Diagram to show a more complex view of the level of origin of the various vertebrate fossils. Note that the precise time of origin is often not clear and that the jawed fish, amphibia and reptilia all arose within a comparatively short time of each other.

(1942), that suggests that the modern amphibia are diphyletic, the anurans coming from one stock of bony fish whilst the urodeles came from another; the two lines being separate in the Early Devonian.)

The reptiles arose in the Carboniferous. There are certain forms such as *Seymouria* that are of interest in that they have body characters that are reptilian and head characters that are amphibian. *Seymouria* is sometimes thought of as a link between the Amphibia and reptiles. Unfortunately *Seymouria* is found in the Permian whilst the first reptiles arose in the Pennsylvanian, some 20 or so million years earlier. The situation concerning the origin of the mammals is not very much more clear, though the mammals certainly evolved at a later date than the first reptiles. Just how the major groups of the mammals evolved is not very clear. Thus we have three distinct mammalian lines, the Monotremes, the Marsupials and the Placentals, and there is no good evidence that all three came from the same reptilian stock. It is quite possible that many of the mammalian-like characters such as warm-bloodedness, double circulation through the heart, development of the neopallium, development of hair and secretion by milk glands, could be homoiologous and that the mammals resemble the arthropods in that they are not a phylum but a grade of organisation that has been achieved many times from a basic stock.

There are other points of interest that arise when we consider the time of origin of these various groups. On embryological grounds it had been considered that cartilage was more primitive than bone. Thus cartilage appeared in the embryo in most cases before bone did, and the elasmobranch fishes in many ways appeared simpler than the bony fish. The elasmobranches were in fact considered to be more primitive than the bony fish and it was not till more attention was paid to the palaeontological dating that it become clear that the elasmobranches were more recent than the Osteichthyes. The Osteichthyes arose in the Early and Middle Devonian whilst the elasmobranches arose in the Middle–Late Devonian. Furthermore most of the fossil groups were bony when first found but there was a tendency to reduce the ossification so that the later forms are less bony and more cartilaginous. The palaeontological evidence thus indicates that bone is more primitive than cartilage and in this respect conflicts with ideas that are derived from embryological studies.

The fact that the groups Agnatha, Placoderms, Osteichthyes, Chrondrichthyes, Amphibia and Reptiles all arose within a relatively short time of each other (possibly by explosive evolution, the explosion lasting over 50 million years) means that one has to be much more accurate in dating the fossils than if it had taken, say, 300 million years.

There are two main ways of dating rocks: an objective method of using radioactive *data* and a subjective method by which one analyses the relative position of the rocks and their included fossils and then comes to conclusions concerning the contemporaneity and priority of the different strata. Neither of these methods is completely free from objection, as we shall now see.

Radioactive dating of rocks

There are various methods by which it is possible to use the ratio of various radioactive materials to determine the absolute age of rocks. Some of these will briefly be mentioned here.

The first method is the uranium–lead method, or the so-called radiogenic lead method. Nier (1939) published a review of this method as applied to various samples and later reviews of the subject have been written by Knopf (1948), Kulp (1955, 1956) and Ahrens (1956). There is also the book by Zeuner (1958) which discusses the various methods of dating material. The common lead isotope has the atomic weight of 204. However, other

isotopes of lead occur and these are formed during the breakdown
of uranium and thorium. Thus

$$U^{238} \longrightarrow Pb^{206}$$
$$U^{235} \longrightarrow Pb^{207}$$
$$Th^{232} \longrightarrow Pb^{208}$$

The rate of decay of $U^{238} \rightarrow Pb^{206}$ is constant and can be determined by experimental observation. It is usually expressed as a
half-life period, i.e. the time taken for x g of uranium to decay
into $x/2$ g of uranium. Since the half-life period of the three
reactions above are known it is possible to get three checks on the
age of any piece of rock that contains U^{235}, U^{238} and Th^{232}. In
less than one million years all three reactions come into equilibrium
and the ratio of the values U^{238}/Pb^{206}; U^{235}/Pb^{207} and Th^{232}/Pb^{208}
should be constant. The ratio of Pb^{207}/Pb^{206} should also be constant since the ratio of U^{235}/U^{238} is constant. This means that on
old rocks all three methods should give results of approximately
the same value.

   This radiogenic dating has been of the greatest value in determining the age of the earth. These studies indicate that the
earth is much older than most people had thought and that it is
of the order of 4,500 million years old. But when the radiogenic
methods are applied to more recent rocks, especially those bearing
fossils, two serious handicaps arise. The first is that this method
can of course only be applied to rocks that contain radiogenic
lead; that is, lead derived from uranium or thorium. These rocks
are usually pegmatites, i.e. rocks formed from the residues after
granite has crystallised out from the liquid mass. This implies
that the material at some stage or another has been molten and
that therefore it is unlikely to contain fossils. Secondly there are
considerable differences in the age as determined from the different ratio of the isotopes 206/207, 206/238, 207/235 or 208/232.
Thus Kulp (1955) has published a table giving *data* for forty-five
different samples of material, the lead ratios being determined by
mass spectrometer; and of these only seven are believed to be
accurate to within 5%. Some are very inaccurate, due, it is
believed, to the loss of radon by diffusion from the rocks in the
series U^{238}/Pb^{206}. Another difficulty is due to the amount of

non-radiogenic lead present in the material. Where this is high there is a corresponding high error in the estimation. This can lead to an error of 700 million years in the exceptional case of the Caribou Mine, Colorado, where the deposit contained as much as 97% lead. The correct age of the deposit was 25 million years old.

From the stratigraphic point of view the radiogenic *data* is a little disappointing. Thus Kulp (1955) states, " The only thoroughly satisfactory sample from a stratigraphic point of view is the Swedish kolm which contains Upper Cambrian fossils." Unfortunately the isotopic ages obtained by Nier for this sample do not agree. Thus:

$$U^{238}/Pb^{206} \text{ gave } 380 \text{ million years}$$
$$U^{235}/Pb^{207} \text{ gave } 440 \text{ million years}$$
$$Pb^{207}/Pb^{206} \text{ gave } 800 \text{ million years}$$

The correct age appears to be 440 million years and it is probable that the other values are in error due to radon loss.

There are two other locations of fossiliferous rocks that have also been accurately radiogenically dated. A pitchblende from Colorado has been dated as 60 million years old. This had been placed at the beginning of the Eocene. It will be remembered that the dating of the Eocene was tentatively done by Matthew (1914) by estimating the time required for the evolution of the horse. Matthew decided that it took 45 million years, i.e. he differed by 15 million years from the radiogenic dating.

All these dates are based on pitchblendes which might have percolated through into the examined strata and they could in fact have been derived from some other strata than that in which they have been discovered. Kulp in discussing the uranium contents of rocks states that the high uranium concentration is often associated with a carbonaceous deposit and it is conceivable that the uranium was accumulated by some biochemical process before the rock systems became molten.

The fact that pegmatites are few and far between makes it improbable that the uranium method will have extensive use in dating fossils. It is more likely that some other method will be of greater use.

Potassium method

This is one of the most promising new methods for the dating of rocks. Potassium is one of the most common elements in rock and its isotope K^{40} occurs as 0.0119% of the natural element. K^{40} decays to form Ca^{40} by beta emission, the decay having a half-life period of 1.35×10^9 years. This would take us back 1,000 million years. There is a second path of decay open to K^{40}. It can capture an electron and turn into A^{40}. This latter system has not yet been fully worked out but it is probable that both methods will prove of great use in dating rock strata (Ahrens 1956). A paper by Mayne, Lambert and York (1959) shows that whereas the previous methods estimated the Upper Cambrian to be 450 million years old, the K method gives a value of 650 million years. This would mean that these rocks are some 200 million years older than previously thought, a point of considerable interest since it indicates the size of the errors to be expected in estimates based on other methods.

Subjective Methods

We have, then, as yet, no accurate objective clock that will allow us to determine the absolute age of the majority of the rocks of the world. Instead we have to go mainly on stratigraphical *data* and there too we find several problems.

From a stratigraphical point of view one cannot state the absolute age of a given piece of rock; all that one can do is estimate the relative age of the rock, and where this is based on the thickness of the deposit and the rate of deposition, the results are bound to be only approximate. If the various levels are complex and stratified, then it is often possible to determine contemporaneity, especially when the strata are close together. The situation is much more complex when one has to decide if rocks in different parts of the world are contemporaneous. Thus the great Caledonian Oregony gave rise to a marked separation of the Devonian and Silurian rocks in North-West Europe, but this separation is not found in the Appalachian syncline. Similarly the rock strata in Maryland, U.S.A., show an unbroken series of deposits from the Late Silurian to Early Devonian without any sign of a boundary.

To what extent can one place the fishes found in England, Scandinavia, Germany and the United States in their correct relative temporal positions? To what extent can one assume that the climatic conditions in different parts of the world would not have affected the distribution of animals so that a local change in climatic conditions lasting some millions of years did not lead to a migration of specimens into different parts of the world? For then, if one decided that strata with similar fossils were of similar age, one could be in serious error. What is the maximum error that can be allowed in the estimation of the date of any given series of rocks? All these questions have to be answered before one can decide that fish A is earlier than fish B in time.

There is another and in this case a minor difficulty. This concerns the nature and naming of the strata. There is often some discussion as to whether a given series of rocks should be placed at the bottom of one level or the top of another. Thus the Tremadoc has been placed both at the top of the Cambrian and also at the bottom of the Ordovician. Another problem concerns the Downtonian; are these Late Silurian or Early Devonian rocks? This point is only of importance when the fossils are classified as Late Silurian or Early Devonian instead of Downtonian. It is indeed a great tribute to the work of the geologists and palaeontologists that so much agreement has been reached concerning the dating of the various strata. But it is unfortunate that the difficulties are often glossed over and only the most simple story presented. When one is dealing with the evolution of the basic vertebrate types all within a comparatively short time of each other, the problem of accurate dating becomes one of critical importance. Many of the conclusions that we have today are only tentative ones.

We can state with certainty that the earliest bony fragments are those of Agnathan fish. These are separated by some 300 million years from the earliest mammalian fragments. But it is much more difficult to decide just how much earlier the Agnatha were than the very first Placoderms; or how much earlier the first Osteichthyes were than the first elasmobranches, or when the first Amphibia arose relative to the time of origin of the reptiles. We can believe that one group arose before the other and there is good evidence that one group of fossils may be commonly found before another, but when it comes down to giving a precise date, or even a reasonable estimation of the time of origin of the groups, it is quite another matter. Thus we don't know the time or the source of origin of the vertebrates. We do not know the relationship between the Agnatha and the Placoderms. We do not know the ancestry of the Osteichthyes or Chondrichthyes. We do not know if the Amphibia are monophyletic or diphyletic. We do not know if the mammals are monophyletic or polyphyletic.

In spite of the ignorance on these basic points certain changes have been taking place. Thus in older classifications one could read about a group called " Pisces." This group is no longer considered a suitable classificatory unit and it has been broken down into the groups Agnatha, Placoderms, Osteichthyes and so on. In other words the group " Pisces " was a complex grade of organisation, a grade corresponding to the " fish level of complexity." Further studies may show that the Amphibia, reptiles and mammals are all grades of organisation and not necessarily very closely related groups of animals.

RATES OF EVOLUTION

Within recent years the study of genera and species in the various geological strata has been put on a quantitative basis. Thus if one studies a group of animals and examines the number of genera present, say, in the Ordovician and then examines how many of these genera are present in more recent strata one can make a calculation of the time over which each genus existed and from this one may come to certain tentative conclusions about the evolution of the groups as a whole.

This technique, together with various others, has been applied with considerable success by G. G. Simpson in his books *Tempo and Mode in Evolution* (1944) and *The Major Features of Evolution* (1953). These works satisfy two desires in the reader: the first is for an intelligent approach to fossil animals and the realisation that they once were living animals; the second is for a treatment of evolution and palaeontology in a mathematical and symbolic manner.

Simpson takes two groups, the Lamellibranches and the Carnivores (excluding the Pinnipedia) and for each draws up a table showing the number of genera present in the Ordovician, Silurian, Devonian and so on, and also the level at which each

genus disappeared. From this information it is possible to draw
a curve showing the percentage of Devonian genera that are alive
at more recent times. From these curves one can see that the
lamellibranches differs from the Carnivora in that the mean
survivorship for a lamellibranch genus is some 78 million years
whilst that for the Carnivora is 6½ million years. Simpson concludes, " The *data* undoubtedly exaggerate the difference for
various reasons, but it is safe to say that the carnivores have
evolved, on the average, some ten times as fast as pelecypods
(lamellibranches) " (1944). This does not mean that all lamellibranch genera lasted for 78 million years and hence ten times
as long as each carnivore genus; the values referred to are mean
values.

Simpson's views concerning the mean length of genera have
not gone unchallenged. In particular Williams (1957) has made
some interesting objections to the techniques employed by
Simpson. Williams states, " An ever-increasing number of papers
dealing with the development of fossil groups contains a host of
graphical and numerical devices designed to provide a sober tone
of objectivity to the accompanying text. On the whole they appear
to be extremely useful but there is a real danger that the student
will lose sight of the tenuous and arbitrary nature of most of the
data used in the compilation of such charts, for there is always a
tendency to accept numbers as the only worthwhile facts in papers
of this kind."

Williams goes on to point out that a great deal in such calculations depends on the nature of the systematics of the groups studied
and whether the systematists working on the groups were
" lumpers " or " splitters." The former group as many species
as possible together into one genus, the latter separate each species
into a separate genus! If a lumper has been at work on a group,
there would be few genera and each would exist for a long period
of geological time. There are also comparatively few genera if a
group has not been " monographed " for some time. (There
appears to be a good correlation between the number of monographs that has been published on a group and the number of
genera described for such a group (Cooper and Williams 1952).)
When the concept of " lumpers and splitters " is applied to the
Brachiopods, a group that includes *Lingula* which has remained

unchanged since the Early Cambrian and which Simpson classifies
as a slowly evolving group, it becomes clear that different views
concerning the survivorship of the genera can be obtained by
examining the genera described in 1894, 1929 or 1956; the mean
life of a genus being 64, 56 or 53 million years respectively. This
would still mean little when compared to the 6½ million years
of the carnivore genus, i.e. the brachiopods would appear to have
evolved some seven times more slowly.

Williams then points out that the figure for the carnivores is
that of a small group taken over the climax of their evolutionary
history, i.e. when the carnivores show the greatest variation and
formation of new genera. On the other hand the figures for the
lamellibranches and the brachiopods are taken over the whole
range of the animals' geological record and both groups are
found from almost the earliest geological time. Williams therefore
suggests that it is more logical to consider the climax of evolution
of such group as the brachiopods. This would be the 231 genera
of Ordovician times. The average duration of these genera is 16
million years and some have as short a duration as 10 million years.
Thus if the climax of evolution is taken for both the Carnivora and
Brachiopoda, the difference of the mean duration of each genus
changes from one of 6½ million years and 53 million years to one of
6½ million years and 16 million years. It would be interesting to
have the similar calculations applied to the lamellibranches and,
say, the early reptiles, the latter showing an explosive type of
evolution that took place some time ago, thus allowing the post-
climax period to be analysed. The example just quoted concerning
the rate of evolution of the Brachiopoda shows how careful one
must be in assuming that conclusions are valid unless one makes
a careful consideration and analysis of the *data* supporting
these conclusions.

THE EVOLUTION OF THE HORSE

It would not be fitting in discussing the implications of Evolution to leave the evolution of the horse out of the discussion. The
evolution of the horse provides one of the keystones in the teach-
ing of evolutionary doctrine, though the actual story depends to a
large extent upon who is telling it and when the story is being

told. In fact one could easily discuss the evolution of the story of the evolution of the horse.

It started when Kowalevsky in 1874 working with European and Asian forms drew up the scheme shown below.

Equus (Pleistocene-Recent)

↑

Hipparion (Pliocene)

↑

Anchitherium (Miocene)

↑

Palaeotherium (Eocene)

These fossil types showed the trends in the evolution of the modern horse, i.e. increase in size of the body, reduction in the number of digits, molarisation of the premolars, etc, even though in fact later workers showed that *Palaeotherium*, *Anchitherium* and *Hipparion* were not even on the main line to the evolution of the horse. In particular, Kowalevsky was handicapped in studying only Old World horses whilst it has been clearly shown by the magnificent work of American palaeontologists such as Marsh, Cope, Leidy, Osborn and Matthew that the major development of the horse took place in the New World. In 1917 Lull published a scheme showing the then current concept of the evolution of the horse (Fig. 42).

Further research showed that the situation was even more complex than that illustrated by Lull and in 1951 the scheme shown in Fig. 43 was a more accurate account of the evolution of the horse; it will be noted that instead of a simple direct line the pattern has become more and more branched.

To the interested non-specialist there are several things that are puzzling in the accounts of the evolution of the horse. In the first place it is difficult to find a critical account of the basic information. The accounts given by Piveteau (1958) and by Matthew (1926) are more concerned with the names of the intermediate forms and the basic trends of evolution. The account given by Simpson (1951) is of great interest and very readable but it is written for a wider audience. It is necessary to go back to the references given in Matthew (1926) or the papers of Matthew and Stirton (1930),

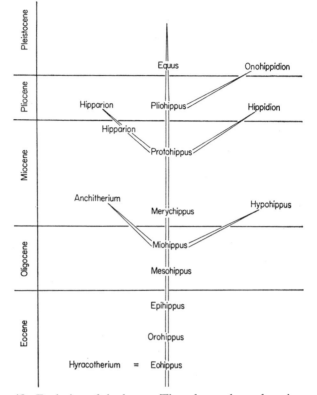

FIG. 42. Evolution of the horse. The scheme shown here is more complex than that suggested by Kowalevsky. (From Lull 1918.)

and Osborn (1905, 1918) to get any satisfaction concerning the fossils themselves.

What does the sceptical reader hope to find out? It takes a great deal of reading to find out for any particular genus just how complete the various parts of the body are and how much in the illustrated figures is due to clever reconstruction. The early papers were always careful to indicate by dotted lines or lack of shading the precise limits of the reconstructions, but later authors are not so careful. Secondly it is difficult to find out just how many specimens of a given genus are available for study. Thus it is one thing to know that our information on *Hyracotherium* is based on, say, 500 specimens, and another if our information is

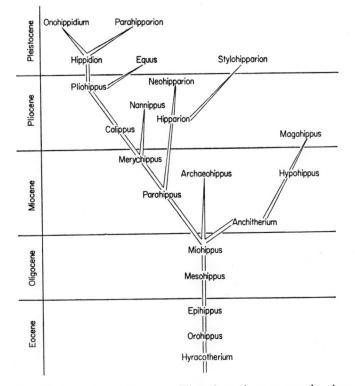

FIG. 43. Evolution of the horse. The scheme is more complex than that suggested by Lull. (After Simpson 1951.)

based on five specimens. In the former case we have a very good idea of the form of the genus and the extent to which its characters overlap those of related genera (that is, provided that the 500 specimens are not just isolated cones of teeth!). A certain amount of information concerning the number of fossil horses is available. Thus Simpson (1943) quotes the numbers of specimens of fossil horses in the American Museum of Natural History as follows: Lower Eocene 397; Middle Eocene 54; Upper Eocene 11; Lower Oligocene 30; Middle Oligocene 125; Upper Oligocene 39. The same author in his account of horses (1951) has an appendix " where to see fossil horses." He mentions that there are some fifty-two mounted skeletons of fossil horses in the

11—IOE

U.S.A. and probably a total of 100 in the world. There are not any mounted skeletons of *Eohippus, Archaeohippus, Megahippus, Stylohipparion, Nannippus Calippus, Onohippidium* or *Parahippus*, and none in the United States of *Anchitherium* or *Hipparion*. There are, however, several thousand of horse fragments collected in the various museums of the world. It is expecting a great deal to have fully prepared specimens of all the major genera of fossil horses. But since the horse is such a key example in the evolutionary doctrine it is important that our knowledge of the fragments be collected, possibly in the first place as a card index system and then later published as a catalogue, so that the results can be made available in synoptic form to all those interested.

A third problem concerns the validity of the various genera and generic differences. The number of genera described has increased considerably. Thus Kowalevsky in 1874 knew of three; Lull in 1917 described fifteen; Simpson in 1945 lists twenty-six genera. To some extent this is due to the discovery and description of new material but one wonders how valid these genera really are.

Another problem concerns the dating of these genera. When Matthew worked out the time taken for the evolution of the horse, he came to the conclusion it would take some 45 million years. His calculation was a rough one but it provided a useful guide. Since then this calculation has been modified by a uranium dating which places the Eocene back to 60 million years. Over this 60 million years we have had some twenty-six genera and a large number of species of fossil horse evolving and it would be of the greatest interest to know the relative positions of these animals to one another, together with some indication as to the accuracy of the relative dating. Thus if we could know the parts of, say, *Mesohippus* skeleton that have been found in perfect condition, the number of specimens, fragments and so on of *Mesohippus* that are available, the strata from which each of these was derived, the degree of contemporaneity of the strata plus or minus so many million years, then we should have no qualms in accepting the evidence presented to us. At present, however, it is a matter of faith that the textbook pictures are true, or even that they are the best representations of the truth that are available to us at the present time.

One thing concerning the evolution of the horse has become clear. The story of the evolution of the horse has become more and more complex as further material is collected, and instead of a simple family tree the branches of the tree have increased in size and complexity till the shape is now more like a bush than a tree. In some ways it looks as if the pattern of horse evolution might be even as chaotic as that proposed by Osborn (1937, 1943) for the evolution of the Proboscidea, where, " in almost no instance is any known form considered to be a descendant from any other known form; every subordinate grouping is assumed to have sprung, quite separately and usually without any known intermediate stage, from hypothetical common ancestors in the Early Eocene or Late Cretaceous " (Romer 1949). We now know that the evolution of the horse did not always take a simple path. In the first place it is not clear that *Hyracotherium* was the ancestral horse. Thus Simpson (1945) states, " Matthew has shown and insisted that *Hyracotherium* (including *Eohippus*) is so primitive that it is not much more definitely equid than tapirid, rhinocerotid, etc, but it is customary to place it at the root of the equid group."

Similarly it is clear that though in general the horses did increase in size, certain genera such as *Orohippus*, *Archaeohippus* and *Nannippus* appear to have been smaller than their ancestors. Edinger (1948) from her studies of the casts of the skull and the brains of fossil horses has concluded that the brain surface of the early fossil horses was perfectly smooth and that the sulci have developed at a later date. This would indicate that any resemblances that have been drawn between the sulci on the brain of the modern horse and those of other mammals are either due to convergent evolution or to homoiology.

It is quite likely that further studies will show that the complexity of horse evolution will prove to be as great as that found in the Proboscidea, Rhinocerotidea or Camelidae.

CHAPTER 10

CONCLUSIONS

WHAT conclusions, then, can one come to concerning the validity of the various implications of the theory of evolution? If we go back to our initial assumptions it will be seen that the evidence is still lacking for most of them.

(1) The first assumption was that non-living things gave rise to living material. This is still just an assumption. It is conceivable that living material might have suddenly appeared on this world in some peculiar manner, say from another planet, but this then raises the question, " Where did life originate on that planet? " We could say that life has always existed, but such an explanation is not a very satisfactory one. Instead, the explanation that non-living things could have given rise to complex systems having the properties of living things is generally more acceptable to most scientists. There is, however, little evidence in favour of biogenesis and as yet we have no indication that it can be performed. There are many schemes by which biogenesis could have occurred but these are still suggestive schemes and nothing more. They may indicate experiments that can be performed, but they tell us nothing about what actually happened some 1,000 million years ago. It is therefore a matter of faith on the part of the biologist that biogenesis did occur and he can choose whatever method of biogenesis happens to suit him personally; the evidence for what did happen is not available.

(2) The second assumption was that biogenesis occurred only once. This again is a matter for belief rather than proof. It is convenient to believe that all living systems have the same fundamental chemical processes at work within them, but as has already been mentioned, only a few representatives from the wide range of living forms have so far been examined and even

these have not been exhaustively analysed. From our limited experience it is clear that the biochemical systems within protoplasm are not uniform, i.e. there is no established biochemical unity. Thus we are aware that there are systems other than the Embden–Meyerhof and the tricarboxylic cycles for the systematic degradation of carbohydrates; a total of six alternative methods being currently available. High-energy compounds other than those of phosphorus have been described; the number of vital amino-acids has gone up from twenty to over seventy; all these facts indicate that the biochemical systems may be very variable. The morphological systems in protoplasm, too, show considerable variation. It is possible that some aspects of cell structure such as the mitochondria and the microsomes might have arisen independently on several distinct occasions. It is also probable that two or more independent systems have evolved for the separation of chromosomes during cell division.

It is a convenient assumption that life arose only once and that all present-day living things are derived from this unique experience, but because a theory is convenient or simple it does not mean that it is necessarily correct. If the simplest theory was always correct we should still be with the four basic elements—earth, air, fire and water! The simplest explanation is not always the right one even in biology.

(3) The third assumption was that Viruses, Bacteria, Protozoa and the higher animals were all interrelated. It seems from the available evidence that Viruses and Bacteria are complex groups both of which contain a wide range of morphological and physiological forms. Both groups could have been formed from diverse sources so that the Viruses and Bacteria would then be an assembly of forms that contain both primitive and secondarily simplified units. They would each correspond to a Grade rather than a Subkingdom or Phylum. We have as yet no definite evidence about the way in which the Viruses, Bacteria or Protozoa are interrelated.

(4) The fourth assumption was that the Protozoa gave rise to the Metazoa. This is an interesting assumption and various schemes have been proposed to show just how the change could have taken place. On the other hand equally interesting schemes have been suggested to show the way in which the Metaphyta

could have given rise to both the Protozoa and the Metazoa. Here again nothing definite is known. We can believe that any one of these views is better than any other according to the relative importance that we accord to the various pieces of evidence.

(5) The fifth assumption was that the various invertebrate phyla are interrelated. If biogenesis occurred many times in the past and the Metazoa developed on several finite occasions then we might expect to find various isolated groups of invertebrates. If on the other hand biogenesis was a unique occurrence it should not be too difficult to show some relationship between all the various invertebrate phyla.

It should be remembered, for example, that though there are similarities between the cleavage patterns of the eggs of various invertebrates these might only reflect the action of physical laws acting on a restrained fluid system such as we see in the growth of soap bubbles and not necessarily indicate any fundamental phylogenetic relationship.

As has already been described, it is difficult to tell which are the most primitive from amongst the Porifera, Mesozoa, Coelenterata, Ctenophora or Platyhelminthia and it is not possible to decide the precise interrelationship of these groups. The higher invertebrates are equally difficult to relate. Though the concept of the Protostomia and the Deuterostomia is a useful one, the basic evidence that separates these two groups is not as clear cut as might be desired. Furthermore there are various groups such as the Brachiopoda, Chaetognatha, Ectoprocta and Phoronidea that have properties that lie between the Protostomia and the Deuterostomia. It is worth paying serious attention to the concept that the invertebrates are polyphyletic, there being more than one line coming up to the primitive metazoan condition. It is extremely likely that the Porifera are on one such side line and it is conceivable that there could have been others which have since died away leaving their progeny isolated; in this way one could explain the position of the nematodes. The number of ways of achieving a specific form or habit is limited and resemblances may be due to the course of convergence over the period of many millions of years. The evidence, then, for the affinities of the majority of the invertebrates is tenuous and circumstantial; not the type of evidence that would allow one to form a verdict of definite relationships.

(6) The sixth assumption, that the invertebrates gave rise to the vertebrates, has not been discussed in this book. There are several good reviews on this subject. Thus Neal and Rand (1939) provide a useful and interesting account of the various views that have been suggested to explain the relationship between the invertebrates and the vertebrates. The vertebrates have been derived from the annelids, arthropods, nemerteans, hemichordates and the urochordates. More recently Berrill (1955) has given a detailed account of the mode of origin of the vertebrates from the urochordates in which the sessile ascidian is considered the basic form. On the other hand, almost as good a case can be made to show that the ascidian tadpole is the basic form and that it gave rise to the sessile ascidian on the one hand and the chordates on the other. Here again it is a matter of belief which way the evidence happens to point. As Berrill states, "in a sense this account is science fiction."

(7) We are on somewhat stronger ground with the seventh assumption that the fish, amphibia, reptiles, birds and mammals are interrelated. There is the fossil evidence to help us here, though many of the key transitions are not well documented and we have as yet to obtain a satisfactory objective method of dating the fossils. The dating is of the utmost importance, for until we find a reliable method of dating the fossils we shall not be able to tell if the first amphibians arose after the first choanichthian or whether the first reptile arose from the first amphibian. The evidence that we have at present is insufficient to allow us to decide the answer to these problems.

One thing that does seem reasonably clear is that many of the groups such as the Amphibia (Save Soderberg 1934), Reptilia (Goodrich 1916) and Mammalia appear to be polyphyletic grades of organisation. Even within the mammals there is the suggestion that some of the orders might be polyphyletic. Thus Kleinenberg (1959) has suggested that the Cetacea are diphyletic, the Odontoceti and the Mysticeti being derived from separate terrestrial stocks. (Other groups that appear to be polyphyletic are the Viruses, Bacteria, Protozoa, Arthropoda (Tiegs and Manton 1958), and it is possible that close study will show that the Annelida and Protochordata are grades too.)

In effect, much of the evolution of the major groups of animals has to be taken on trust. There is a certain amount of circumstantial evidence but much of it can be argued either way. Where, then, can we find more definite evidence for evolution? Such evidence will be found in the study of modern living forms. It will be remembered that Darwin called his book *The Origin of Species* not *The Origin of Phyla* and it is in the origin and study of the species that we find the most definite evidence for the evolution and changing of form. Thus to take a specific example, the Herring Gull, *Larus argentatus*, does not interbreed with the Lesser Black-backed Gull, *Larus fuscus*, in Western Europe, the two being separate species. But if we trace *L. argentatus* across the northern hemisphere through North America, Eastern Siberia and Western Siberia we find that in Western Siberia there is a form of *L. argentatus* that will interbreed with *L. fuscus*. We have here an example of a ring species in which the members at the ends of the ring will not interbreed whilst those in the middle can. The separation of what was possibly one species has been going on for some time (in this case it is suggested since the Ice Age). We have of course to decide that this is a case of one species splitting into two and not of two species merging into one, but this decision is aided by the study of other examples such as those of small mammals isolated on islands, or the development of melanic forms in moths. Details of the various types of speciation can be found in the books by Mayr, *Systematics and the Origin of Species* (1942), and Dobzhansky, *Genetics and the Origin of Species* (1951).

It might be suggested that if it is possible to show that the present-day forms are changing and the evolution is occurring at this level, why can't one extrapolate and say that this in effect has led to the changes we have seen right from the Viruses to the Mammals? Of course one can say that the small observable changes in modern species may be the sort of thing that lead to all the major changes, but what right have we to make such an extrapolation? We may feel that this is the answer to the problem, but is it a satisfactory answer? A blind acceptance of such a view may in fact be the closing of our eyes to as yet undiscovered factors which may remain undiscovered for many years if we believe that the answer has already been found.

It seems at times as if many of our modern writers on evolution have had their views by some sort of revelation and they base their opinions on the evolution of life, from the simplest form to the complex, entirely on the nature of specific and intra-specific evolution. It is possible that this type of evolution can explain many of the present-day phenomena, but it is possible and indeed probable that many as yet unknown systems remain to be discovered and it is premature, not to say arrogant, on our part if we make any dogmatic assertion as to the mode of evolution of the major branches of the animal kingdom.

Perhaps it is appropriate here to quote a remark made by D'Arcy Thompson in his book *On Growth and Form*. " If a tiny foraminiferan shell, a *Lagena* for instance, be found living today, and a shell indistinguishable from it to the eye be found fossil in the Chalk or some still more remote geological formation, the assumption is deemed legitimate that the species has ' survived ' and has handed down its minute specific character or characters from generation to generation unchanged for untold millions of years. If the ancient forms be like rather than identical with the recent, we still assume an unbroken descent, accompanied by hereditary transmission of common characters and progressive variations. And if two identical forms be discovered at the ends of the earth, still (with slight reservation on the score of possible ' homoplasy ') we build a hypothesis on this fact of identity, taking it for granted that the two appertain to a common stock, whose dispersal in space must somehow be accounted for, its route traced, its epoch determined and its causes discussed or discovered. In short, the Naturalist admits no exception to the rule that a natural classification can only be a genealogical one, nor ever doubts that ' the fact that we are able to classify organisms at all in accordance with the structural characteristics which they present is due to their being related by descent.' "

What alternative system can we use if we are not to assume that all animals can be arranged in a genealogical manner? The alternative is to indicate that there are many gaps and failures in our present system and that we must realise their existence. It may be distressing for some readers to discover that so much in zoology is open to doubt, but this in effect indicates the vast amount of work that remains to be done. In many courses the student is obliged to read, assimilate and remember a vast amount of factual information on the quite false assumption that knowledge is the accumulation of facts. There seems so much to be learnt that the only consolation the student has is that those who come after him will have even more to learn, for more will be known. But this is not really so; much of what we learn today are only half truths or less and the students of tomorrow will not be bothered by many of the phlogistons that now torment our brains.

It is in the interpretation and understanding of the factual information and not the factual information itself that the true interest lies. Information must precede interpretation, and it is often difficult to see the factual *data* in perspective. If one reads an account of the history of biology such as that presented by Nordenskiold (1920) or Singer (1950) it sometimes appears that our predecessors had a much easier task to discover things than we do today. All that they had to do was realise, say, that oxygen was necessary for respiration, or that bacteria could cause septicaemia or that the pancreas was a ductless gland that secreted insulin. The ideas were simple; they just required the thought and the experimental evidence! Let us have no doubt in our minds that in twenty years or so time we shall look back on many of today's problems and make similar observations. Everything will seem simple and straightforward once it has been explained. Why then cannot we see some of these solutions now? There are many partial answers to this question. One is that often an incorrect idea or fact is accepted and takes the place of the correct one. An incorrect view can in this way successfully displace the correct view for many years and it requires very careful analysis and much experimental *data* to overthrow an accepted but incorrect theory. Most students become acquainted with many of the current concepts in biology whilst still at school and at an age when most people are, on the whole, uncritical. Then when they come to study the subject in more detail, they have in their minds several half truths and misconceptions which tend to prevent them from coming to a fresh appraisal of the situation. In addition, with a uniform pattern of education most students tend to have the same sort of educational background and so in conversation and discussion they accept common fallacies and agree on matters based on these fallacies.

It would seem a good principle to encourage the study of " scientific heresies." There is always the danger that a reader might be seduced by one of these heresies but the danger is neither as great nor as serious as the danger of having scientists brought up in a type of mental strait-jacket or of taking them so quickly through a subject that they have no time to analyse and digest the material they have " studied." A careful perusal of the heresies will also indicate the facts in favour of the currently accepted doctrines, and if the evidence against a theory is over-whelming and if there is no other satisfactory theory to take its place we shall just have to say that we do not yet know the answer.

There is a theory which states that many living animals can be observed over the course of time to undergo changes so that new species are formed. This can be called the " Special Theory of Evolution " and can be demonstrated in certain cases by experiments. On the other hand there is the theory that all the living forms in the world have arisen from a single source which itself came from an inorganic form. This theory can be called the " General Theory of Evolution " and the evidence that supports it is not sufficiently strong to allow us to consider it as anything more than a working hypothesis. It is not clear whether the changes that bring about speciation are of the same nature as those that brought about the development of new phyla. The answer will be found by future experimental work and not by dogmatic assertions that the General Theory of Evolution must be correct because there is nothing else that will satisfactorily take its place.

BIBLIOGRAPHY

AHRENS, L. H. (1956) Radioactive methods for determining geological age. *Rep. Progr. Phys.* **19**; 80.

AMANO, S. (1957) Structure of Centrioles and spindle body as observed under electron and phase contrast microscope; a new extension fibre theory concerning mitotic mechanisms in animal cells. *Cytologia.* **22**; 193.

BAKER, J. R. (1948) The status of the Protozoa. *Nature, Lond.* **161**; 548 and 587.

BALDWIN, E. (1940) *An Introduction to Comparative Biochemistry.* Cambridge University Press.

BALDWIN, E. (1957) *The Dynamic Aspects of Biochemistry.* 3rd edition. Cambridge University Press.

BALDWIN, E., and NEEDHAM, D. M. (1937) A contribution to the comparative biochemistry of muscular and electrical tissues. *Proc. roy. Soc.* B**122**; 197.

BALDWIN, E., and YUDKIN, W. H. (1948) Phosphagen in annelids (Polychaeta). *Biol. Bull.* **95**; 273.

BALDWIN, E., and YUDKIN, W. H. (1949) The annelid phosphagen with a note on phosphagen in Echinodermata and Protochordata. *Proc. roy. Soc.* B**136**; 614.

BALFOUR, F. M. (1880) *Comparative Embryology.* Macmillan, London.

BEATTY, I. M., MARGRATH, D. I., and ENNOR, A. H. (1959) Biochemistry of Lombricine. *Nature, Lond.* **183**; 591.

de BEER, G. (1954) The evolution of the Metazoa. In *Evolution as a Process.* (Edited by Huxley, J., Hardy, A. C., and Ford, E. B.) p. 24. Allen and Unwin, London.

de BEER, G. (1958) *Embryos and Ancestors* (3rd Ed.). Oxford University Press.

van BENEDEN, E. (1876) Recherches sur les dicyemides, survivants actuels d'un embranchement de mesozoaires. *Bull. Acad. Belg. Cl. Sci.* 2nd ser. **41**; 1160; **42**; 35.

BERGMAN, W. (1944) The sterols of starfish. *J. org. Chem.* **9**; 281.

BERGMAN, W. (1949) Comparative biochemical studies of the lipids of marine invertebrates with special reference to the sterols. *J. Mar. Res.* **8**; 137.

BERGMAN, W., and LOW, E. M. (1947) Remarks concerning the structure of sterols from marine invertebrates. *J. org. Chem.* **12**; 67.

BERGMAN, W., MCLEAN, M. J., and LESTER, D. (1943) Sterols from various marine invertebrates (Echinoids). *J. org. Chem.* **8**; 271.

BERNAL, J. D. (1954) The origin of life. *New Biology* **16**; 28. Penguin Books, London.

BERRILL, N. J. (1955) *The Origin of the Vertebrates.* Oxford University Press.

BONNER, J. T. (1949) The demonstration of acrasin in the later stages of the development of the slime mold *Dictyostelium discoideum. J. exp. Zool.* **110**; 259.

BOYDEN, A. (1953) Comparative Evolution with special references to primitive mechanisms. *Evolution.* **7**; 21.

BRESSLAU, E. (1933) Turbellaria. In *Handbuch der Zoologie.* (Edited by Kukenthal, W., and Krumbach, T.) de Gruyler, Berlin. Vol. 2, part 1; p. 52.

BUNTING, M. (1926) Studies on the life cycle of *Tetramitus rostratus. J. Morph.* **42**; 23.

BUTSCHLI, O. (1880-89) Protozoa. In *Bronns Klassen und Ordnungen des Thierreichs.* Berlin.

CARLGREN, O. (1925) Die Tetraplatien. Wiss Ergeb Deut Tiefsee Exped. *Valdivia.* **19**.

CAULLERY, M. (1952) *Parasitism and Symbiosis.* (Translated by Lysaght, A. M.) Sidgwick and Jackson, London.

CHATTON, E. (1920) Les Péridiniens parasites, morphologie, reproduction, ethologie. *Arch. Zool. exp. gén.* **59**; 1.

CHARGAFF, E. (1957) On nucleic acids and nucleoproteins. *Harvey Lect.* **52**; 57.

CLAUS, F. W. (1887) *Lehrbuch der Zoologie.* Elenvert, Marburg.

COHEN, S. S. (1955a) Other pathways of carbohydrate metabolism. In *Chemical Pathways in Metabolism.* (Edited by Greenberg, D. M.) Vol. 1; pp. 173–233. Academic Press.

COHEN, S. S. (1955b) Comparative biochemistry and virology. *Adv. Vir. Res.* **3**; 1.

COOPER, G. A., and WILLIAMS, A. (1952) Significance of the stratigraphic distribution of the Brachiopods. *J. Paleont.* **26**; 326.

CUÉNOT, L. (1952) Phylogene due Regne Animal, in *Traité de Zoologie.* (Edited by Grassé, P. P.) Masson et Cie.

DAWYDOFF, C. (1928) *Traité d'embryologie comparée des invertebres.* Masson et Cie.

DELAGE, Y. (1898) Sur la place des Spongiaires dans la classification. *C.R. Acad. Sci. Paris.* **136**; 545.

DELAGE, Y. (1898) On the position of sponges in the animal kingdom. 4th *Int. Cong. Zool.* p. 57.

DELAGE, Y., and HÉROUARD, E. (1896) La Cellule et Les Protozoaires. In *Traité de Zoologie Concrete.* Schliecher, Paris.

DOBZHANSKY, T. G. (1951) *Genetics and the Origin of Species.* Columbia University Press.

DODSON, E. O. (1956) A note on the systematic position of the Mesozoa. *Syst. Zool.* **5**; 37.

DOFLEIN, F. (1916) *Lehrbuch der Protozoenkunde.* Fischer, Jena.

DOGIEL, V. (1906) *Haplozoon armatum,* de vertreter einer neuen Mezozoangruppe. *Zool. Anz.* **30.**

DUBOSCQ, O., and GRASSÉ, P. P. (1933) L'appareil parabasal des Flagelles. *Arch. Zool. exp. gén.* **73**; 381.

DUBOSCQ, O., and TUZET, O. (1937) L'ovogénèse, la fécondation et les premier stades du dévelopement des éponges calcaires. *Arch. Zool. exp. gén.* **79**; 157.

EDINGER, T. (1948) Évolution of the horse brain. *Geol. Soc. Amer. Mem.* **25.**

EGGLETON, P., and EGGLETON, G. P. (1928) Further observations on phosphagen. *J. Physiol.* **65**; 15.

ENNOR, A. H., and MORRISON, J. F. (1958) Biochemistry of the phosphagens and related guanidines. *Phys. Rev.* **38**; 631.

FAURÉ-FREMIET, E. (1930) Growth and differentiation in the colonies of *Zoothamnion alternans. Biol. Bull.* **58**; 28.

FELL, H. B. (1948) Echinoderm embryology and the origin of the chordates. *Biol. Rev.* **23**; 81.

FRAENKEL-CONRAT, H., and WILLIAMS, R. C. (1955) Restitution of active tobacco mosaic virus from its inactive protein and nucleic acid components. *Proc. nat. Acad. Sci. Wash.* **41**; 690.

FRANZ, V. (1924) *Geschichte der Organismen.*

FRUTON, J. S., and SIMMONS, S. (1958) *General Biochemistry.* Wiley, New York.

FRY, B. A., and PEEL, J. L. (1954) *Autotrophic Bacteria. Symp. Soc. gen. Microbiol.* Cambridge.

GOETTE, A. (1902) *Lehrbuch der Zoologie.* Voss, Leipzig.

GOODRICH, E. S. (1916) On the classification of the Reptilia. *Proc. roy. Soc.* B**89**; 261.

von GRAFF, L. (1904–8) Acoela und Rhabdocoelida. in *Bronns Klassen und Ordnungen des Thierreichs.*

GRASSÉ, P. P. (1952) *Traité de Zoologie.* Vol. 1; Fascicle 1. Masson et Cie.

GRAY, J. (1931) *Experimental Cytology.* Cambridge University Press.

GRIFFITHS, D. E., MORRISON, J. F., and ENNOR, A. H. (1957) The distribution of guanidines, phosphagens and n-amidino phosphokinases in Echinoids. *Biochem. J.* **65**; 612.

GROBBEN, K. (1908) Die systematische Einteilung des Tierreiches. *Verh. zool. bot. Ges. Wien.* **58**; 491.

GRÖNDTVED, J. (1956) Taxonomical studies in some Danish coastal localities. Meddel. Danmark. *Fiskeri-og-Havundersogelse.* New Series. **1**; No. 12.

GURWITSCH, A. (1926) *Das Problem der Zellteilung physiologisch betrachtet.* Springer, Berlin.

HADŽI, J. (1944) Turbelarijska Teorija Knidarijev. *Razpr. Slovensk. Akad. Mat. Prirod., Ljubljana.* **3**; 1. (Summary in German. pp. 190–239.)

HADŽI, J. (1951) Ali imajo ktenofore lastne ozigalke? *Razpr. Slov. Akad. Znan. Umetn.* **4**; 13.

HADŽI, J. (1953) An attempt to reconstruct the system of animal classification. *Syst. Zool.* **2**; 145.

HAECKEL, E. (1870) Biologische Studien. I: Studien über die moneren und andere Protisten, nebst einer Rede über Entwickelungsgang und Aufgabe der Zoologie. Jena.

HAECKEL, E. (1872) Die Kalkschwamme (Calcispongien oder Grantien) Ein Monographie in zwei Banden Text und einem Atlas mit 60 Tafeln Abildungen. I: Genereller Theil; II: Specieller Theil; III: Illustrativer Theil.

HAECKEL, E. (1877) *Biologische Studien.* II: Studien zur Gastrae. Theorie. Jena.

HAECKEL, E. (1899) *Naturliche Schopfungs-Geschichte.* George Reimer, Berlin.

HALDANE, J. B. S. (1954) The Origins of Life. *New Biology.* **16**; 12. Penguin Books, London.

HAND, C. (1955) A study of the structure, affinities and distribution of *Tetraplatia volitans.* Busch. *Pacif. Sci.* **9**; 332.

HARDY, A. C. (1953) On the origin of the metazoa. *Quart. J. Micr. Sci.* **94**; 441.

HARPER, H. (1959) *Review of Physiological Chemistry.* Lange. Medical Publ., California.

HEATH, H. (1928) A sexually mature Turbellarian resembling Müller's larva. *J. Morph.* **45**; 187.

HEDLEY, R. D. (1958) The biology and cytology of *Haliphysema. Proc. zool. Soc. Lond.* **130**; 569.

HERSHEY, A. D. (1956) Bacteriophage T2. Parasite or organelle? *Harvey Lect.* **51**; 229.

HOBSON, G. E., and REES, K. R. (1957) The annelid phosphagens. *Biochem. J.* **65**; 305.

HOLLAENDER, A., and SCHOEFFEL, E. (1931) Mitogenetic rays. *Quart. Rev. Biol.* **6**; 215.

HUXLEY, T. H. (1849) Memoir of the Anatomy and affinities of the Medusae. *Phil. Trans. Roy. Soc. Lond.*

HUXLEY, T. H. (1891) *A Manual of the Anatomy of the Invertebrated Animals.* Churchill, London.

HYMAN, L. H. (1940) *The Invertebrates.* Vol. I. Protozoa through Ctenophora. McGraw-Hill, New York.

HYMAN, L. H. (1955) *The Invertebrates.* Vol. IV. Echinodermata. McGraw-Hill, New York.

IVANOV, A. V. (1954) New Pogonophora from the Eastern Seas. (Translated by Petrunkewitch, A.) *Syst. Zool.* **3**; 68.

IVANOV, A. V. (1955) The main features of the organization of the Pogonophora. (Translated by Petrunkewitch, A.) *Syst. Zool.* **4**; 170.

IVANOV, A. V. (1956) On the systematic position of the Pogonophora (Translated by Petrunkewitch, A.) *Syst. Zool.* **5**; 165.

IVANOV, A. V. (1957) Neue Pogonophora aue dem nordwestlichen Teil des Stillen Ozeans. *Zool. Jb.* **85**; 430.

JAGERSTEN, G. (1955) On the early phylogeny of the Metazoa. The Bilaterogastraea theory. *Zool. Bidr. Uppsala.* **30**; 321.

JARVIK, E. (1942) On the structure of the snout of Crossopterygians and lower Gnathostomes in general. *Zool. Bidr. Uppsala.* **27**; 235.

JOHNSTON, G. (1838) *A History of British Zoophytes.* Edinburgh.

KEILIN, D., and WANG, Y. L. (1945) Haemoglobin in the root modules of Leguminous plants. *Nature, Lond.* **155**; 223.

KLEBS, G. (1892) Flagellatenstudien. *Zeit. wiss. Zool.* **55**; 265.

KLEINENBERG, S. E. (1959) On the origin of the Cetacea. *Proc. XV International Congr. Zool.* p. 445.

KNIGHT, C. A. (1949) Constituents of viruses. *Ann. Rev. Microbiol.* **3**; 121.

KNOPF, A. (1948) Time in earth history. Pages 1–9 in *Palaeontology, Genetic and Evolution.* (Edited by Jepsen, G., Simpson, G. G., and Mayr, E.) Princeton University Press.

KOMAI, T. (1922) Studies on two aberrant ctenophores, *Coeloplana* and *Gastrodes.* Published by the author. Kyoto, Japan.

KOMAI, T. (1942) The nematocysts in the Ctenophore *Euchlora rubra. Proc. imp. Acad. Tokyo.* **18.**

KOMAI, T. (1951) The nematocysts in the Ctenophore *Euchlora rubra. Amer. Nat.* **85**; 73.

KORNBERG, H. L. (1958) The metabolism of C_2 compounds in microorganisms. *Biochem. J.* **68**; 535.

KOWALEWSKII, V. O. (1842) Sur *l'Anchiterium aurelianeuse* et sur l'histoire paléontologique des Chevaux. *Mem. Acad. imp. Sci.* St. Pet. 7. Vol. 20.

KREBS, H. (1948) The tricarboxylic acid cycle. *Harvey Lect.* **44**; 165.

KREBS, H. (1954) The tricarboxylic acid cycle. In *Chemical Pathways in Metabolism.* (Edited by Greenberg, D.) Vol. 1, p. 109.

KRUMBACH, T. (1927) Scyphozoa. In *Handbuch der Zoologie.* (Edited by Kukenthal, W., and Krumbach, T.) Vol. 1. de Gruyter, Berlin.

KUKENTHAL, W., and KRUMBACH, T. (1923) *Handbuch der Zoologie.* de Gruyter, Berlin.

KULP, J. L. (1955a) Isotopic dating of the geologic times scale. Geol. Soc. Amer. Special Paper **62**; 609.

KULP, J. L. (1955b) Geological chronometry by radioactive methods. *Adv. Geophys.* **2**; 179. Academic Press.

LANG, A. (1884) Die Polycladen des Golfes con Neapel. *Fauna and flora of Gulf of Naples.* 11.

LANKESTER, E. RAY. (1879) The structure of *Haliphysema tumanowiczii*. *Quart. J. Micr. Sci.* **19.** (New Series); 476.

LANKESTER, E. RAY. (1890) *Zoological Articles Contributed to the Encyclopaedia Britannica.* Adam and Charles Black, London.

LANKESTER, E. RAY. (1900) *A Treatise on Zoology.* Adam and Charles Black, London.

LANKESTER, E. RAY. (1909) Introduction and Protozoa. Fascicle 1. In *A Treatise on Zoology.* Adam and Charles Black, London.

LEDERBERG, J. T. (1947) Gene recombination and linked segregations in *Escherichia coli. Genetics.* **32;** 505.

LEDERBERG, J. T., and TATUM, E. L. (1954) Sex in bacteria; genetic studies. Page 12 in *Sex in microorganisms.* (Edited by Wenrich, D. H., Lewis, I. F., and Raper, J. R.) Amer. Ass. Adv. Sci.

LEUCHTENBERGER, C. (1958) Quantitative determination of DNA in cells by Feulgen microspectrophotometry. *General Cytochemical Methods.* Vol. 1; p. 219. Academic Press.

LIPMANN, F. (1958) Biological sulfur, activation and transfer. *Science,* **128;** 575.

LULL, R. (1917) *Organic Evolution.* Methuen, London.

LURIA, S. E. (1953) Origin and nature of viruses. *General Virology,* Chapter 18. Wiley.

LWOFF, A. (1944) *L'Evolution Physiologique.* Etude des pertes de fonctions chez les microorganismes. Hermann, Paris.

LYNEN, F. (1952) Acetyl Co-A and the fatty acid cycle. *Harvey Lect.* **48;** 210.

LYTTELTON, R. (1951) *The Mysterious Universe.* Hodder and Stoughton, London.

MADISON, K. M. (1953) The organism and its origin. *Evolution.* **7;** 211.

MANTON, S. M. (1948) Studies on the Onychophora. VII: the early embryonic stages of *Peripatopsis* and some general considerations concerning the morphology and phylogeny of the Arthropoda. *Phil. Trans.* B**233;** 483.

MARCUS, E. (1958) On the evolution of the animal phyla. *Quart. Rev. Biol.* **33;** 24.

MARKHAM, R., SMITH, K. M., and LEA, D. (1942) The sizes of viruses and the methods employed in their estimation. *Parasitology.* **34;** 315.

MARKHAM, R., and SMITH, J. D. (1951) Chromatographic studies of nucleic acids. *Biochem. J.* **49;** 401.

MATTHEW, W. D. (1914) Time ratios in the evolution of mammalian phyla; a contribution to the problem of the age of the earth. *Science.* **40;** 232.

MATTHEW, W. D. (1926) The evolution of the horse. *Quart. Rev. Biol.* **1;** 130.

MATTHEW, W. D., and STIRTON, R. A. (1930) Equidae from the Pliocene of Texas. Univ. Calif. Pub. *Bull. Dept. Geol.* **29;** 349.

MAYNE, K. I., LAMBERT, R. ST. J., and YORK, D. (1959) The geological time scale. *Nature, Lond.* **183;** 212.

MAYR, E. (1942) *Systematics and the Origin of Species.* Columbia University Press.

McCONNAUGHEY, B. H. (1951) The life cycles of the dicyemid Mesozoa. *Univ. Calif. Publ. Zool.* **55;** 1.

MEISTER, A. (1957) *The Biochemistry of Amino-Acids.* Academic Press.

MEYERHOF, O. (1928) Uber die verbreitung der arginin-phosphorsaure in der muskelature der wirbellosen. *Arch. Sci. biol. Napl.* **12;** 536.

MOORE, R. C. (1955) Invertebrates and the geologic time scale. *Geol. Soc. Amer.* Special Papers **62;** 547.

MOORE, R. C. (1956) *Treatise of Invertebrate Paleontology.* Geol. Soc. Amer. Kansas University Press.

MORTENSEN, T. (1912) A sessile ctenophore. *Tjalfiella tristoma* and its bearing on Phylogeny. *Brit. Ass. Adv. Sci.* Sect. D.

MOSER, F. (1925) Siphonophora in *Handbuch der Zoologie.* (Edited by Kukenthal, W., and Krumbach, T.) Vol. 1. de Gruyter, Berlin.

NEAL, H. V., and RAND, H. W. (1943) *Comparative Anatomy.* Blakiston.

NEEDHAM, D. M., NEEDHAM, J., BALDWIN, E., and YUDKIN, J. (1932) A comparative study of the phosphagens with some remarks on the origin of vertebrates. *Proc. roy. Soc.* B110; 260.

NEEDHAM, J. (1931) *Chemical Embryology.* Cambridge University Press.

NIER, A. O., THOMPSON, R. W., and MURPHY, B. F. (1941) The isotopic constitution of lead and the measurement of the geological time scale. *Phys. Rev.* **60;** 112.

NORDENSKIOLD, E. (1920) *History of Biology.* Tudor Pub. Co.

O'KANE, D. J. (1941) The synthesis of riboflavine by staphylococci. *J. Bact.* **41;** 441.

OPARIN, A. I. (1957) *The Origin of Life on the Earth.* (3rd Ed.) Oliver and Boyd, London.

OSBORN, H. F. (1905) Ten years' progress in the mammalian palaeontology of North America. *C.R. 6th Internat. Congr. Berne.* Page 86.

OSBORN, H. F. (1918) Equidae of the Oligocene, Miocene, and Pliocene of North America. *Mem. Amer. Mus. nat. Hist.* **2;** 1.

OSBORN, H. F. (1936, 1942) *Proboscidea.* A monograph of the discovery, evolution, migration and extinction of the mastodonts and elephants of the world. Vol.1: Moeritherioidea, Deinotherioidea, Mastodontoidea. Vol. 2: Stegodontoidea, Elephantoidea. Amer. Mus. nat. Hist.

OUSDAL, A. P. (1956) *Syst. Zool.* **5;** 161.

PALEY, W. (1802) *Evidences of Christianity.* Johnson.

PANTIN, C. F. A. (1942) The excitation of nematocysts. *J. exp. Biol.* **19;** 294.

PICARD, J. (1955) Les nematocystes du Ctenaire *Euchlora rubra. Rec. Trav. Stat. Mar. Endoume.* **15;** 99.

PIRIE, N. W. (1954) On making and recognising life. *New Biology.* **16;** 41. Penguin Books, London.

PIVETEAU, J. (1958) *Traité de Palaeontologie.* Masson et Cie.

POCOCK, M. A. (1933) Volvox in South Africa. *Ann. S. Afr. Mus.* **16**; 523.

PRINGLE, J. W. S. (1953) The origin of life. *Soc. exp. Biol. Symp.* **7**; 1. Cambridge.

PRINGLE, J. W. S. (1954) The evolution of living matter. *New Biology.* **16**; 54. Penguin Books.

PRINGSHEIM, E. G. (1948) Taxonomic problems in the Euglenineae. *Biol. Rev.* **23**; 46.

PRINGSHEIM, E. G., and HOVASSE, R. (1950) Les relations de parente entre Astasiees et Euglenacees. *Arch. Zool. exp. gén.* **86**; 499.

PROSSER, C. L. (Editor). (1950) *Comparative Animal Physiology.* Saunders.

RADL, E. (1930) *The History of Biological Theories.* (Translated by Hatfield, E. J.) Oxford University Press.

RALPH, P. M. (1959) Notes on the species of pteromedusan genus *Tetraplatia* Busch 1851. *J. Mar. Biol. Ass.* **38**; 369.

RASMONT, R., BOUILLON, J., CASTIAUX, P., and VENDERMEERSSCHE, G. (1958) Ultra structure of the choanocyte collar cells in fresh water sponges. *Nature, Lond.* **181**; 58.

RAVEN, CHR. P. (1958) *Morphogenesis: the Analysis of Molluscan Development.* Pergamon Press.

REES, K. R. (1958) Personal communication.

REICH, W. (1948) *The Discovery of the Orgone.* I: The function of the orgasm. II: The Cancer biopathy. Orgone Inst. Press., New York.

REMANE, A. (1954) Die Geschichte der Tiere. In *Die Evolution der Organismen.* (Edited by Heberer, G.) Vol. 2; p. 340. G. Fisher, Jena.

ROBIN, Y., van THOAI, N.-G., and PRADEL, L.-A. (1957) Sur une nouvelle guanidine monosubstituée biologique; L'Hirudonine. *Biochem. Biophys. Acta.* **24**; 381.

ROBINOW, C. F. (1946) Chapter in *The Bacterial Cell.* (Edited by Dubos, R. J.) Harvard University Press.

ROCHE, J., and ROBIN, Y. (1954) Sur les phosphagens des Éponges. *C.R. Soc. Biol.* **148**; 541.

ROCHE, J., van THOAI, N.-G., and ROBIN, Y. (1957) Sur la présence de creatine chez les invertébrés et as signification biologique. *Biochem. Biophys. Acta.* **24**; 514.

ROMER, A. S. (1949) Time series and trends in animal evolution. In *Genetics, Palaeontology, and Evolution.* (Edited by Jepson, G., Simpson, G., and Mayr, E.) Princeton University Press.

SAVE SODERBERGH, G. (1934) Some points of view concerning the evolution of the vertebrates and the classification of this group. *Arkiv. Zool.* **26**; 1.

SAVILLE KENT, W. (1878) The foraminiferal nature of *Haliphysema tumanowiczii* demonstrated. *Ann. mag. nat. Hist. Sci.* V. **2**; 68.

SAVILLE KENT, W. (1880) *A Manual of the Infusoria.* Bogue, London. 3 vols.

SCHRADER, F. (1953) *Mitosis.* (2nd ed.) Columbia University Press.

SEAMAN, G. R. (1952) The phosphagen of protozoa. *Biochem. Biophys. Acta.* **9**; 693.

SEDGWICK, A. (1884) On the nature of the metameric segmentation and some other morphological questions. *Quart. J. Micr. Sci.* **24**; 43.

SEDGWICK, A. (1915) *A Student's Textbook of Zoology.* Vol. 3. Sonnenschein.

SHAPLEY, H. (1958) *Of Stars and Men.* Beacon Press, New York.

SIMPSON, G. G. (1944) *Tempo and Mode in Evolution.* Columbia University Press.

SIMPSON, G. G. (1945) The principles of classification and a classification of mammals. *Bull. Amer. Mus. nat. Hist.* **85**; 1.

SIMPSON, G. G. (1951) *Horses.* Oxford University Press, New York.

SIMPSON, G. G. (1953) *The Major Features of Evolution.* Columbia University Press.

SINGER, C. (1950) *A History of Biology.* Schumann.

SPOONER, E. T. C., and STOCKER, B. A. D. (1956) Bacterial anatomy. *Symp. Soc. gen. Microbiol.* Cambridge.

STIRTON, R. A. (1940) Phylogeny of North American Equidae. *Univ. Calif. Publ. Bull. Geol.* **25**; 165.

STUNKARD, H. W. (1954) The life history and systematic relations of the Mesozoa. *Quart. Rev. Biol.* **29**; 230.

SUMMERS, E. (1938) Some aspects of normal development in colonial ciliate *Zoothamnion alternans. Biol. Bull.* **74**; 117.

SUMMERS, E. (1938) Form regulation in *Zoothamnion alternans. Biol. Bull.* **74**; 130.

SWANN, M. (1951) Protoplasmic structure and mitosis. *J. exp. Biol.* **28**; 417.

TAKAHASHI, W. N., and ISHII, M. (1953) A macromolecular protein associated with tobacco mosaic virus infection; its isolation and properties. *Amer. J. Bot.* **40**; 85.

van THOAI, N.-G., ROCHE, J., ROBIN, Y., and THIEM, N. (1953) Sur deux nouvaeux phosphagenes; la phosphotaurocyamine et la phosphoglycocyamine. *C. R. Biol. Soc. Paris.* **147**; 1241.

van THOAI, N.-G., and ROBIN Y. (1951) Les methylases chez les invertebres marins. *C. R. Biol. Soc. Paris,* **145**; 1674.

van THOAI, N.-G., ROCHE, J., and ROBIN, Y. (1957) Sur la presence de creatine chez les invertebres et la significance biologique. *Biochem. Biophys. Acta.* **24**; 514.

van THOAI, N.-G., and ROBIN, Y. (1954) Metabolism des derives guanidyles. Sur un nouvelle guanidine monosubstituée biologique. L'ester guanido ethyl seryl phosphorique (lombrincine) et le phosphagen correspondant. *Biochem. Biophys. Acta.* **14**; 76.

THOMPSON, D'ARCY, W. (1942) *On Growth and Form.* Cambridge University Press.

TIEGS, O. and MANTON, S. M. (1958) The evolution of the Arthropoda. *Biol. Rev.* **33**; 255.

TUZET, O. (1945) Sur les aggregats de choanocytes et la question de la *Proterospongia* avec quelques remarques sur les reseux formé par les cellules dissociées. *Archiv. Zool. exp. gén.* **84**; 225.

ULRICH, W. (1950) Vorschlage zu einer Revision der Grosseinteilung des Tierreiches. *Verh. dtsch. zool. Ges. Zool. Anz. Suppl.* **15**; 244.

VERBINSKAYA, N. A., BORSUK, V. N., and KREPS, E. (1935) Biochemistry of muscle contraction in *Cucumaria frondosa. Arch. Sci. Biol.* Moscow. **38**; 369.

WHEELER, L. R. (1939) *Vitalism.* Witherby.

WHITE, E. I. (1948) The vertebrate fauna of the lower old red sandstone of the Welsh Border. *Bull. Brit. Mus. nat. Hist. Geol.* **1**; 51.

WILLIAMS, A. (1957) Evolutionary rates of Brachiopods. *Geol. Mag.* **94**; 201.

WILLMER, E. N. (1956) Factors which influence the acquisition of flagella by the amoeba, *Naegleria gruberi. J. exp. Biol.* **33**; 583.

WILLMER, E. N. (1958) Further observations on the " metaplasia " of an amoeba, *Naegleria gruberi. J. Embryol. exp. Morph.* **6**; 187.

ZEUNER, F. (1958) *Dating the Past.* Methuen.

ZINSSER, H. (1957) *Zinsser's Bacteriology.* 11th edition. (Edited by Smith, D. T., and Connant, D. F.) Appleton-Century-Crofts.

AUTHOR INDEX

SUBJECT INDEX

Pleodorina, 37, 40
Pleurobrachia, 76, 84, 86, 89, 116, 117, 119
Podarke, 102
Pogonophora, 56, 83, 106
Polycelis, 116, 117, 119
Polycladida, resemblances to Cteno-phora, 90–94
Polykrikos, 40
Polymeria, 110
Polyp, 79, 80, 83, 96, 97
Polyphyletic, 13, 152, 153
Pomatoceros, 102
Porifera, 50, 54–71, 82, 132, 152
Porpita, 79
Potassium, 140
Poterion, 56
Primitiveness, 51, 54
Proboscidea, 149
Protanthea, 82
Proterospongia, 57
Protostomia, 102–105
Protozoa, origin, 24–25, 46
    colonial forms, 36–44, 47, 57 *et seq.*
    interrelationship, 32–35
    most primitive, 26, 47
    syncytia, 44–47
Protista, 13
Pterobranchiata, 83
Radial symmetry, 52, 96
Radioactive dating of rocks, 137–140
Rates of evolution, 142–144
Rhinoscerotoidea, 149
Rhizoflagellata, 33
Rhopalura, 72, 75
Rickettsia, 21–22
Sabella, 54, 123, 127
Sabellaria, 116, 117, 119, 123
Saccocirrus, 102
Sacculina, 51
Sakaguchi's test, 120, 121
Salpa, 87
Sappinia, 46
Scolecida, 102
Scyphozoa, 68, 78, 79, 80, 81, 82, 84, 87, 93–96
Sepia, 117, 118, 119, 120, 126
Seymouria, 136
Simplicity, 51
Siphonophora, 76
Sipunculus, 114, 117, 118, 127

Special theory of Evolution, 157
Spaerechinus, 121, 126
Spiral cleavage, 92, 105
Spirochona, 27
Spirographis, 116, 117, 123
Sponges, origin, 57, 71
    specialised characters, 55–56
    simple characters, 55
Spongiaria, 94
Sporozoa, 26, 43, 44, 45, 53
Stellasterol, 130, 131
Sterols, 129–133
    Porifera, 132
    Echinodermata, 129, 133
Stichopus, 114
Stronglylocentrotus, 117, 119, 124
Stylotella, 132, 133
Suberites, 132
Sulphur, 12, 23
Symmetry, 52–54, 90, *see also* Radial symmetry; Bilateral symmetry
Synapta, 117, 119
Syncytia, 39, 44, 46, 63
Taurocyamine, 125
Taurocyamine phosphate, 125–128
Teredo, 103
Tetrahymena, 127, 128
Tetramitus, 31
Tetraplatia, 87, 89
Tentaculata, 107
Thetia, 126
Thiobacillus, 11, 23
Thorium, 138
Trachylina, 78, 84
Trematodes, 72, 94
Tricarboxylic acid cycle, 9, 10, 11
Trichonympha, 26
Trochosphaera, 54
Turbellaria, 83, 89–94, 108
Tubularia, 76, 77
Uranium, 137–139
Velella, 79
Vermes, 35
Vertebrata, 102, 113, 129, 134 *et seq.*, 153
Viruses, 18–21, 151
Volvox, 37, 39, 40, 44, 46, 57, 59, 60–61
William of Occam, 9
Zoöthamnion, 41, 42, 48